From the Collection of
Friends of the Columbus
Metropolitan Library

CRABS

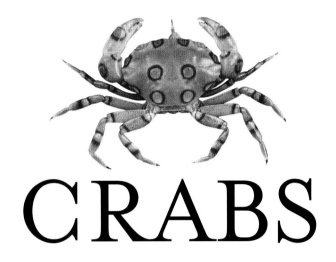

CRABS

A GLOBAL NATURAL HISTORY

PETER J.F. DAVIE

PRINCETON UNIVERSITY PRESS
PRINCETON AND OXFORD

Dedicated to my much-loved wife, Kathleen, who completed me, but was taken from us too early, as well as to our wonderful children, Nicholas and Johanna. As Head Librarian at the Queensland Museum for 30 years, Kath was not only my best friend but also gave me enormous professional support in my crab journey.

Published by Princeton University Press
41 William Street, Princeton, New Jersey 08540
6 Oxford Street, Woodstock, Oxfordshire OX20 1TR
press.princeton.edu

Text copyright © 2021 Peter Davie

Design and layout copyright © 2021 by Quarto Publishing plc

All rights reserved. No part of this publication may be reproduced or transmitted in any form, or by any means, electronic or mechanical, including photocopying, recording or by any information storage-and-retrieval system, without written permission from the copyright holder.

Some photographic images in this work were taken by officers and/or employees of the US Government as part of their official duties and are not copyrightable.

Library of Congress Control Number 2021930494
ISBN: 978-0-691-20171-9
Ebook ISBN: 978-0-691-23013-9

Typeset in Chaparral Pro and Futura

Printed and bound in Singapore

This book was conceived,
designed and produced by
The Bright Press,
part of the Quarto Group
The Old Brewery, 6 Blundell Street,
London N7 9BH, England

Publisher JAMES EVANS
Editorial Director ISHEETA MUSTAFI
Managing Editor JACQUI SAYERS
Art Director and Cover Design JAMES LAWRENCE
Development Editor ABBIE SHARMAN
Project Manager ANGELA KOO
Design JC LANAWAY
Illustrations JOHN WOODCOCK
Picture Research PETER DAVIE & ALISON STEVENS

PAGE 2 IMAGE: The sea cucumber crab (*Lissocarcinus orbicularis*) is only ever found living in symbiosis with sea cucumbers. It lives on the surface of the host, near the oral tentacles and anus, where there is a plentiful food supply. It will also crawl in through the anus to hide from predators.

CONTENTS

Introducing crabs 6
Classification chart 18

1. EVOLUTIONARY PATHWAYS 20

What's in a name? Crab precursors and impostors 22
Crabs in prehistory 30
Evolutionary trends within brachyurans 34
    Robber crab 44
    Halloween hermit crab 46
    Puget Sound king crab 48
    Durian crab 50
    Slender frog crab 52
    Shaggy shore crab 54
    Green-spotted gall crab 56
    Spiny-clawed deep-sea crab 58

2. ANATOMY AND PHYSIOLOGY 60

How crabs are put together: external anatomy 62
What's hidden inside: internal anatomy and physiology 76
    Yellowline arrow crab 84
    Slender-clawed boxer crab 86
    Gaudy clown crab 88
    Lopsided crab 90
    Pretty crested reef crab 92
    Rough-shelled porter crab 94
    Guinot's agile reef crab 96
    Malaysian face-stripe mangrove crab 98
    Horn-eyed ghost crab 100

3. CRAB ECOLOGY 102

A most successful group… 104
Crab environments 112
Living with others: symbioses 120
    Christmas Island blind cave crab 124
    Gaimard's spider crab 126
    Hourdezi's hydrothermal vent crab 128
    Sally Lightfoot crab 130
    Red-kneed soldier crab 132
    Lewinsohn's sponge crab 134
    Harlequin swimming crab 136
    Sculptured crab 138
    Adams zebra crab 140

4. REPRODUCTION, COGNITION AND BEHAVIOUR 142

Reproduction 144
Behaviour and intelligence 156
    Striped box crab 162
    Christmas Island red crab 164
    Candy crab 166
    Superb decorator crab 168
    Arrowhead crab 170
    Stalk-eyed shore crab 172
    Thick-legged fiddler crab 174
    Garfunkel's crab 176

5. EXPLOITATION AND CONSERVATION 178

Crabs as food 180
Crabs and disease 190
Crabs behaving badly 194
Crab conservation 198
    Red king crab 202
    Spanner crab 204
    Tiger crab 206
    Common European spider crab 208
    Horsehair crab 210
    Chinese mitten crab 212
    European edible crab 214
    Little red vampire crab 216

Glossary / Further Reading 218
Index 219
Picture Credits 223
Acknowledgements 224

INTRODUCING CRABS

CRABS ARE TRULY CHARISMATIC ANIMALS. They are thought to have first emerged as a separate group from other Crustacea during the early Jurassic, around 180 million years ago (mya), and so were witness to the reign of the dinosaurs. Evidence is strong that crabs had already left the sea and entered freshwater rivers, lakes and swamps by 135 mya, and were no doubt relished as food by the smaller saurians – as they still are today by many birds and reptiles.

The word 'crab' conjures up many images, with seafood being predominant amongst them. Around 1.5 million tonnes of 'true' crabs (see page 14) are consumed worldwide every year. Only about 14 species are involved in the main commercial industry, but many more are eaten by indigenous peoples and those in poorer communities. Even quite small crabs, if they are common enough, will not be spared.

OPPOSITE TOP: The spanner or red frog crab (*Ranina ranina*) buries itself in bare sandy areas, where it acts as an ambush predator (see also page 204).

OPPOSITE: The massive claw of this Australian southern giant crab (*Pseudocarcinus gigas*) is the largest known of any brachyuran. It can be as long as an adult human forearm.

ABOVE: The ringed pebble crab (*Leucosia anatum*) belongs to a diverse family, many with smooth, rounded shells. They like to bury themselves with only their snouts emerging from the bottom.

ABOVE RIGHT: The southern kelp crab (*Taliepus nuttallii*) is a common shore crab of the tropical Pacific coast of the Americas—an algae eater, varying in colour from yellow-orange to dark purple or reddish-brown.

While crabs have great commercial and nutritional value, they are much, much more than just seafood. Crabs play critical roles in the healthy ecology of coral reefs, mangrove swamps and shallow coastal waters. Armies of tiny crabs keep our beaches clean, either by ravenous scavenging of anything dead, or by sifting masses of sand through their mouths at every low tide, pulling out the microscopic detritus of animals and plants. And some of the larger crab species, such as the giant mud crab (*Scylla serrata*), can produce up to 6 million eggs per female, so it is not hard to understand how crab larvae have become a crucially important component of the plankton community upon which other marine animals depend.

Most people know very little about the rich diversity of crab shape and habit, or the crucial part they play in sustaining the wellbeing of the planet. This book draws on the latest research breakthroughs in classification, evolution, physiology, ecology and behaviour to provide new insights into the lives of crabs, to which only a few experts are normally privy – presented alongside stunning images that show crabs as you have never seen them before.

WHY ARE CRABS SO FASCINATING?

Crabs, more than any other invertebrate, are firmly embedded in the human psyche. To be 'crabby' (easily annoyed and defensive) is a universal English expression; 'claws' and 'walking sideways' also immediately evoke images of crabs. The origins of the medical term 'cancer' are credited to the Greek physician Hippocrates (460–370 BCE). He coined the terms *carcinos* (tumour) and *carcinoma* (malignant tumour) based on Carcinus, the giant crab of Greek mythology, having observed that the cut surface of such tumours showed 'the veins stretched on all sides as … the crab has its feet'. These terms were grouped under *cancer*, the Latin word for 'crab', by the Roman writer Aulus Cornelius Celsus (c. 25 BCE–c. 50 CE) in his *De Medicina*.

The constellation of Cancer is one of the 12 ancient astrological signs, and likely to have first been recognized by Sumerian stargazers by around 3000 BCE. Sumerians referred to the region's abundant freshwater crabs (the *Potamon* species) as *allul* (literally, 'deceptive digger'), and because Sumer, in southern Mesopotamia, is considered to be the cradle of modern civilization, this is probably the first name ever documented for a crab. It also seems no coincidence that the astrological month of

> *'If we live out our span of life on the earth without ever knowing a crab intimately, we have missed a good friendship.'*
>
> CHARLES WILLIAM BEEBE (1877–1962)
> American naturalist and marine biologist

Cancer occurs in midsummer, from 21 June to 20 July, coinciding exactly with the period when thousands of female *Potamon* crabs emerge from the rivers in that region. They venture into terrestrial habitats to search for protein-rich foods that will allow them to produce eggs rich in yolk, giving their developing young the best chance of survival.

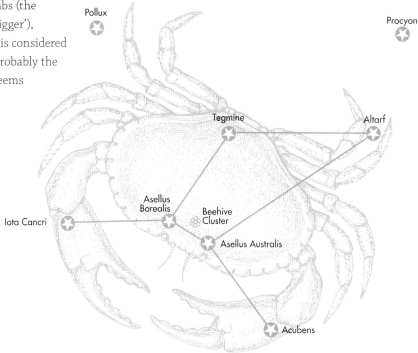

RIGHT: The ancient Sumerians were probably the first to recognize the constellation known as Cancer, and the Babylonians described the *Potamon* species *Nagar-assura* as the 'constellation of the fourth month'.

ABOVE: Mediterranean river crabs (*Potamon fluviatile*) are predators of tadpoles, frogs and fishes. Although known for millennia, and widely distributed through the region, this species is now under threat from pollution, habitat alteration and overfishing.

The ancient Babylonians also had a name for river crabs. An inscribed clay tablet dug from the Euphrates River Valley, dated to 500 BCE, bears the statement, 'the crab called *Nagar-assura* appears as the constellation of the fourth month'. *Nagar-assura* translates as 'workman of the riverbed', a poetic but accurate description of the *Potamon* species that are common throughout the region, and industriously excavate burrows along the banks of rivers, lakes and swamps.

Greek mythology has a different explanation for how Cancer the Crab earned its place in the cosmos. In the course of his second labour, Heracles' vengeful stepmother, the goddess Hera, sent a giant crab to distract Heracles by nipping him, thus giving the multi-headed Hydra a greater chance of defeating him. Heracles dispatched both the crab and the Hydra, but Hera nevertheless rewarded the crab for its loyalty by raising it to a position amongst the stars. A recognizable image of the common Mediterranean river crab (*Potamon fluviatile*) can also be found on ancient coins that were in wide use in the Phoenician and Greek settlements around the Mediterranean. The ancient Greek city-state of Akragas, on the south coast of Sicily, even took the crab as its emblem.

The stars that define Cancer also mark the latitude of the summer solstice, when the Sun reaches its furthest northerly point from the Equator (the Tropic of Cancer), before beginning its return journey south, bringing winter to the northern climes.

Astronomers have continued to name celestial objects in honour of crabs, though perhaps with a little more scientific objectivity and a little less romance! In 1840, William Parsons, 3rd Earl of Rosse, observed a distant stellar object through a small telescope, and made a drawing that looked somewhat crab-like. Thus the violent, fiery supernova death of a distant star in the constellation of Taurus became what is known as the Crab Nebula. In 1967, another crab-like shape was spotted amongst the stars, this time in the Southern Hemisphere – in the constellation Centaurus – so it became the Southern Crab Nebula.

The formal study and description of Crustacea, and crabs, first began with Aristotle. His *Scala Naturae* ('Natural Ladder') was the first system to organize and classify the natural world. He divided animals into two major groups – red-blooded animals (corresponding to vertebrates), and all other creatures without red blood (invertebrates). He then divided the latter into five groups: crustaceans (crabs, lobsters and shrimps); cephalopods (squid, octopus); hard-shelled animals (cockles, trumpet snails); larva-bearing insects (ants, cicadas); and spontaneously generating creatures (sponges, worms). Aristotle's descriptions of individual crustacean species were so accurate that, of the 18 species he discussed, 12 can be identified with certainty as species we know today.

Aristotle's *History of Animals* was written in 350 BCE, and was the foundation upon which classification was based until Carl Linnaeus (the 'father of modern taxonomy') published his *Systema Naturae* nearly 2,100 years later. Linnaeus founded the binomial system of classification (genus–species) that we still use today. '*Cancer* Linnaeus, 1758' has the honour of being one of the oldest official generic names in zoology, but when first used, it was applied to virtually anything 'crustacean', and even to some other aquatic arthropods. Naturalists quickly recognized, however, that *Cancer* was being used too indiscriminately, and by 1802, when the first English edition of Linnaeus's *A General System of Nature* was published, the genus *Cancer* was restricted to a particular kind of brachyuran crab (see page 12).

Much has been learned about crabs since then. The 18 crustaceans described by Aristotle have increased to well over 7,000 crab species alone, and more are being described every year. New techniques – especially the ability to explore genetic relationships – are showing that many seemingly widespread species are really groups of very similar, but separate, species. Such research helps us understand the biodiversity around us, and relate it to the complex history of changes in land connections and ocean basins that have occurred throughout geological time.

OPPOSITE: The aptly named orangutan crab (*Achaeus japonicus*) has the reddish-brown silky hair and long arms of its namesake. It is a tropical Indo-West Pacific species that is often found on bubble corals, or sometimes even anemones, as here.

RIGHT: The Crab Nebula, discovered in 1840, is a supernova in the constellation of Taurus. In 1967, the Southern Crab Nebula was named in the constellation Centaurus.

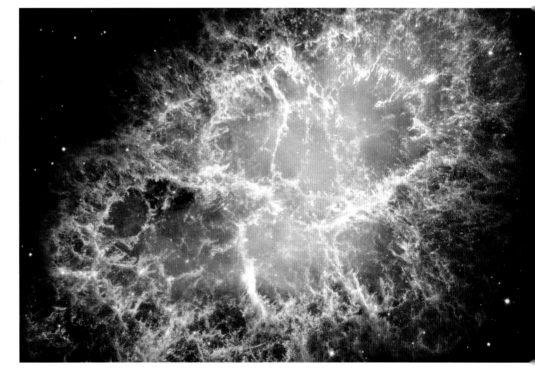

WHAT IS A CRAB?

Crabs are extremely sophisticated crustaceans. Their highly modified and reduced pleon (abdomen or tail) has become tucked neatly away beneath their shell – often locked tightly into place. Indeed, the scientific name for crabs is Brachyura, a term derived from the ancient Greek meaning 'short tail'. This one major innovation allowed crabs to diversify and conquer environments previously largely out of reach of other crustaceans. Not only were the compact bodies of the first crabs free to evolve into a fantastic array of shapes and sizes, but rapid movement in any direction became a reality using specialized limbs. These limbs allowed a range of movement impossible in crustaceans encumbered with a long tail. Long legs have endowed many crabs with great speed, and others with remarkable dexterity. The crab shape was just the catalyst, however, for an extraordinary evolutionary blossoming that involved complex changes to their anatomy, physiology, sensory systems, reproduction and behaviour.

The crab body-plan is arguably more diverse than that of any other decapod group. The giant Japanese spider crab (*Macrocheira kaempferi*) is the largest known arthropod (see page 23) – it can reach 40 cm (16 in) in carapace width, and 19 kg (42 lb) in weight. At the other extreme, a miniature false spider crab, *Neorhynchoplax minima,* is perhaps the tiniest mature crab, with egg-bearing females only 1.4 mm across the carapace. The massive Australian southern giant crab (*Pseudocarcinus gigas*; see pages 6 and 185), weighing in at 17.6 kg (39 lb), comes a close second in size to the Japanese spider crab, but has a wider carapace (46 cm [18 in]), and its massive major chela is the largest known claw of any crab.

Body shapes are enormously variable, ranging from circular to oval, pyriform, pentagonal, hexagonal, trapezoidal or rectangular. Crabs are mostly wider than they are long, but they can also sometimes be much longer than wide.

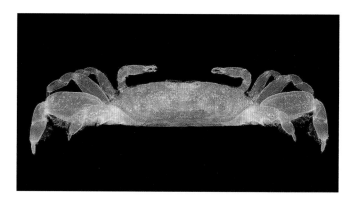

Claws can be massive and powerful, or remarkably small and delicate. Walking legs can be long and stilt-like to raise the body high off the substrate, broad and flat for running, or hook-like for clinging to coral branches. The last pair of legs can be effectively lost (family Hexapodidae), or sometimes vestigial, being present only as a small appendage (Dynomenidae, Palicidae and Retroplumidae). In special cases, the last one or two pairs are held up over the back of the carapace, and possess special nippers to hold sponges or other objects used for camouflage. Body and legs can be smooth and shiny, granular, or armed with an impressive arsenal of protective spines. And many species show marked sexual dimorphism, with males being larger, smaller, or possessing specialized or enlarged claws. In some groups the females are the larger, and males are dwarfed (typically in symbiotic crabs in Pinnotheroidea and Cryptochiroidea).

ABOVE: Members of the pea crab (family Pinnotheridae) are typically commensal with other animals, particularly bivalve molluscs. This wide, slender species of *Pinnixa* shares the narrow tube of a parchment worm.

OPPOSITE: *Cyclocoeloma tuberculata* is one of the 'decorator crabs', which camouflage their shells with a living cover of anemones and other cnidarians (see page 168).

BECOMING CRABBY

Brachyurans have for a long time been divided into two major groupings – 'primitive' crabs (Podotremata) and 'true' crabs (Eubrachyura). However, recent research has split the podotreme families into two evolutionary lines: the Podotremata and the Archaeobrachyura. Podotreme and archaeobrachyuran crabs are much closer in general appearance to a group of 'would-be' crabs known as anomurans (see page 26). This is because their pleon is not as reduced as in eubrachyuran crabs, and (in males) is not locked in place under the crab's body. However, like all real crabs (and unlike all other decapods), they no longer have a 'tail fan' (a central pointed telson surrounded by flattened uropods; see page 63) – only the simple central telson remains as a segment at the end of the pleon. The most significant difference between the two groups comes down to fundamental physical changes in how they reproduce, with a switch from external egg fertilization in Podotremata, to internal fertilization in Eubrachyura (see page 72).

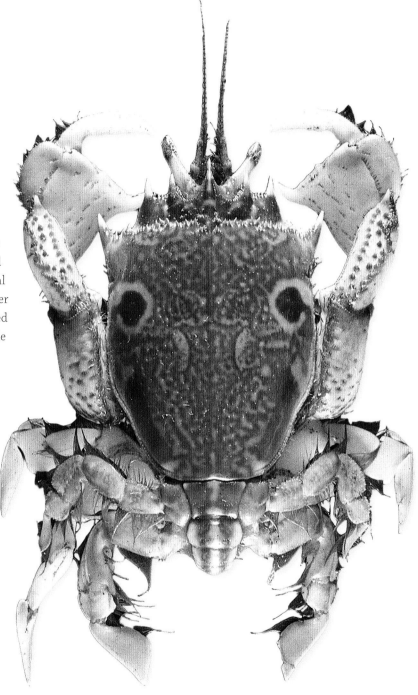

RIGHT: A member of the frog crab family Raninidae, *Notopus dorsipes* is an example of the 'primitive' crabs that first evolved during the Jurassic Period.

Podotreme crabs first appeared at least 180 mya, and over the next 45 million years, podotremes and archaeobrachyurans diversified rapidly into many new forms, dominating the ocean floor unchallenged. But as the Jurassic ended and the early Cretaceous began, the first eubrachyuran crabs appeared, and these were ultimately to radiate into the huge variety of families, genera and species that exist today.

WHERE DO CRABS LIVE?

Crabs have conquered almost every habitat, from the near-freezing blackness of the ocean floor, 4,000 m (13,100 ft) deep, to mountain forests 2,000 m (6,500 ft) above sea level. They can even thrive on the flanks of hydrothermal vents that spew out hot, sulphur-rich water. Some species complete their whole life cycle using only the small pools of water caught in tree holes or the leaf axils of rainforest bromeliads, while others can hibernate in deep clay burrows in hostile desert, waiting six years or more for the rain that will allow their lives to continue. Some groups have even evolved to cope with – and thrive in – the sulphurous anaerobic muds of mangrove swamps, where salinity and temperatures can have dramatic daily fluctuations. Most crabs are still tied to the sea for egg release and larval development, but some large groups have broken this primeval connection, and have direct development through to baby crabs under the wide female pleon, and even a level of maternal care!

ABOVE: The striking mosaic boxer, or pom pom crab (*Lybia tessellata*), carries tiny sea anemones on its claws that it uses for food gathering, defence from predators, and for fighting others of its kind (see also page 86).

LAST WORDS

One of Sir David Attenborough's self-proclaimed greatest television moments was an encounter with Christmas Island red crabs (*Gecarcoidea natalis*; see also pages 200–1), while filming their annual migration in 1990. He described them as:

> '…*a great scarlet curtain moving down the cliffs and rocks towards the sea … it was an astonishing, wonderful sight, but what makes it really stick in the memory is the decision that I should sit in the middle of the beach to deliver my script…. That's how I discovered how difficult it is to deliver lines while several four-inch crabs, each armed with sharp claws, are advancing menacingly up your inner thigh.*'

CLASSIFICATION CHART

- PHYLUM: Arthropoda
- SUBPHYLUM: Crustacea
- ORDER: Decapoda
- INFRAORDER: Brachura

SECTION	SUPERFAMILY	Family
PODOTREMATA	HOMOLODROMIOIDEA	Homolodromiidae
	HOMOLOIDEA	1. Homolidae 2. Latreilliidae 3. Poupiniidae
	DROMIOIDEA	1. Dromiidae 2. Dynomenidae
ARCHAEOBRACHYURA	CYCLODORIPPOIDEA	1. Cyclodorippidae 2. Cymonomidae 3. Phyllotymolinidae
	RANINOIDEA	1. Lyreidiidae 2. Raninidae
EUBRACHYURA Subsection HETEROTREMATA	AETHROIDEA	Aethridae
	BELLIOIDEA	Belliidae
	BYTHOGRAEOIDEA	Bythograeidae
	CALAPPOIDEA	1. Calappidae 2. Matutidae
	CANCROIDEA	1. Atelecyclidae 2. Cancridae
	CARPILIOIDEA	Carpiliidae
	CHEIRAGONOIDEA	Cheiragonidae
	CORYSTOIDEA	Corystidae
	DAIROIDEA	Dairidae

SECTION	SUPERFAMILY	Family
EUBRACHYURA	DORIPPOIDEA	1. Dorippidae 2. Ethusidae
	ERIPHIOIDEA	1. Dacryopilumnidae 2. Eriphiidae 3. Hypothalassiidae 4. Menippidae 5. Oziidae 6. Platyxanthidae
	GECARCINUCOIDEA	Gecarcinucidae
	GONEPLACOIDEA	1. Acidopsidae 2. Chasmocarcinidae 3. Conleyidae 4. Euryplacidae 5. Goneplacidae 6. Litocheiridae 7. Mathildellidae 8. Progeryonidae 9. Scalopidiidae 10. Sotoplacidae 11. Vultocinidae
	HEXAPODOIDEA	Hexapodidae
	HYMENOSOMATOIDEA	Hymenosomatidae
	LEUCOSIOIDEA	1. Iphiculidae 2. Leucosiidae
	MAJOIDEA	1. Epialtidae 2. Inachidae 3. Inachoididae 4. Majidae 5. Mithracidae 6. Oregoniidae
	ORITHYIOIDEA	Orithyiidae
	PALICOIDEA	1. Crossotonotidae 2. Palicidae

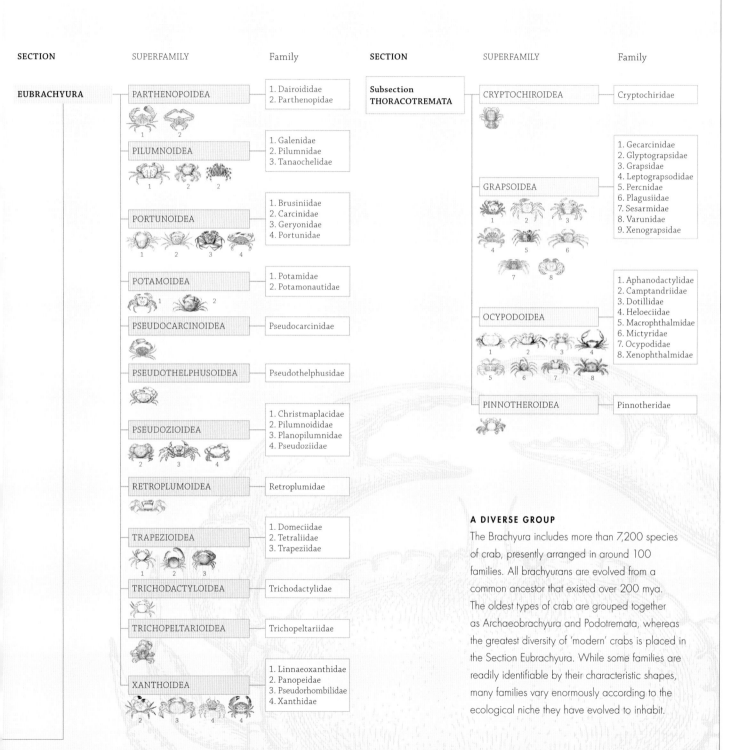

SECTION	SUPERFAMILY	Family
EUBRACHYURA	PARTHENOPOIDEA	1. Dairoididae 2. Parthenopidae
	PILUMNOIDEA	1. Galenidae 2. Pilumnidae 3. Tanaochelidae
	PORTUNOIDEA	1. Brusiniidae 2. Carcinidae 3. Geryonidae 4. Portunidae
	POTAMOIDEA	1. Potamidae 2. Potamonautidae
	PSEUDOCARCINOIDEA	Pseudocarcinidae
	PSEUDOTHELPHUSOIDEA	Pseudothelphusidae
	PSEUDOZIOIDEA	1. Christmaplacidae 2. Pilumnoididae 3. Planopilumnidae 4. Pseudoziidae
	RETROPLUMOIDEA	Retroplumidae
	TRAPEZIOIDEA	1. Domeciidae 2. Tetraliidae 3. Trapeziidae
	TRICHODACTYLOIDEA	Trichodactylidae
	TRICHOPELTARIOIDEA	Trichopeltariidae
	XANTHOIDEA	1. Linnaeoxanthidae 2. Panopeidae 3. Pseudorhombilidae 4. Xanthidae

SECTION	SUPERFAMILY	Family
Subsection THORACOTREMATA	CRYPTOCHIROIDEA	Cryptochiridae
	GRAPSOIDEA	1. Gecarcinidae 2. Glyptograpsidae 3. Grapsidae 4. Leptograpsodidae 5. Percnidae 6. Plagusiidae 7. Sesarmidae 8. Varunidae 9. Xenograpsidae
	OCYPODOIDEA	1. Aphanodactylidae 2. Camptandriidae 3. Dotillidae 4. Heloeciidae 5. Macrophthalmidae 6. Mictyridae 7. Ocypodidae 8. Xenophthalmidae
	PINNOTHEROIDEA	Pinnotheridae

A DIVERSE GROUP

The Brachyura includes more than 7,200 species of crab, presently arranged in around 100 families. All brachyurans are evolved from a common ancestor that existed over 200 mya. The oldest types of crab are grouped together as Archaeobrachyura and Podotremata, whereas the greatest diversity of 'modern' crabs is placed in the Section Eubrachyura. While some families are readily identifiable by their characteristic shapes, many families vary enormously according to the ecological niche they have evolved to inhabit.

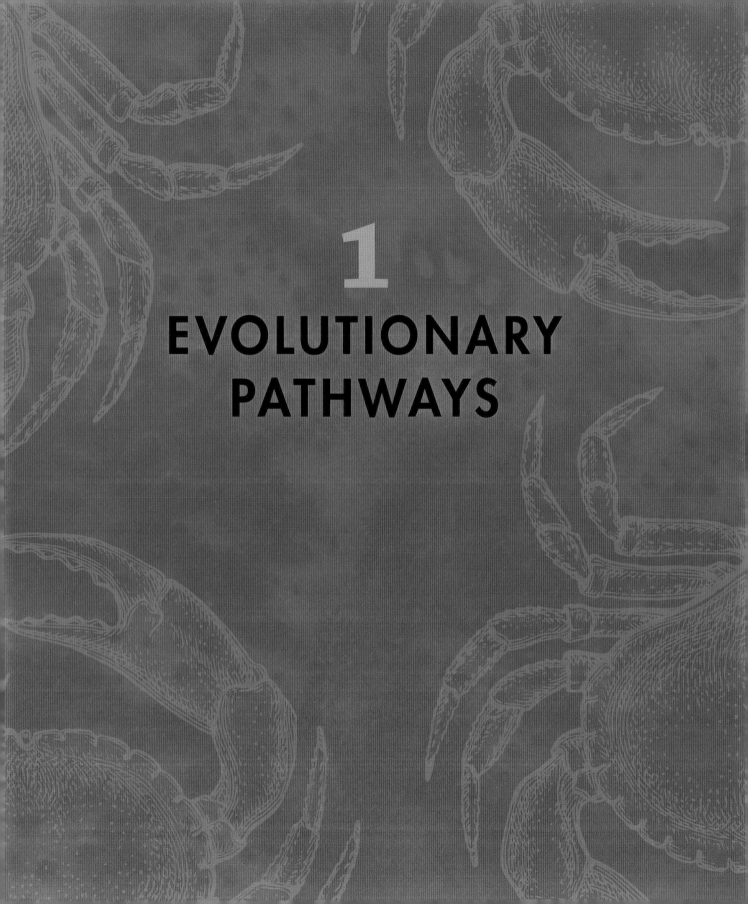

1
EVOLUTIONARY PATHWAYS

WHAT'S IN A NAME? CRAB PRECURSORS AND IMPOSTERS

CRUSTACEANS, INSECTS AND ARACHNIDS are grouped together in the phylum Arthropoda, an amazingly successful group of animals that has an enormous impact on human life. Arthropods comprise more than 80 per cent of all animal life, and more than a million species have been described. 'Arthropoda' is derived from the Greek for 'jointed feet'. It refers to the hinged legs of their hardened exoskeleton – a marvellously versatile structure that variously allows arthropods to walk, run, swim or fly quickly in any direction.

Arthropods first appeared during the great 'Cambrian explosion' around 530 million years ago (mya), when most major animal groups first appear in the fossil record. There has been much debate about how closely different groups of arthropods are related. However, modern research in various fields, including DNA studies, indicates that arthropods are what is called 'monophyletic', meaning that they share one common ancestor.

The living arthropods are divided into four major subgroups, or subphylums: Chelicerata, which includes the arachnids (spiders, mites, ticks), xiphosurans (horseshoe crabs) and pycnogonids (sea spiders); Myriapoda (centipedes and millipedes); Hexapoda (insects); and Crustacea (brine shrimps, remipedes, horseshoe shrimp, barnacles, copepods, fish lice, seed shrimp, and the well-known lobsters, shrimps and crabs). There is also one extinct group well represented in the fossil record, the subphylum Trilobitomorpha (trilobites).

A major genetic study begun in 2011, the 'i5k' project, is intent on documenting and comparing the entire genomes of 5,000 arthropods from across all subphyla. With nearly 10 per cent so far completed, analysis has identified 'families of genes' within each of the major arthropod groups. These gene families are the key to understanding how each group has been able to adapt to the challenges of surviving through half a billion years. The most dynamically changing genes are linked to digestion, chemical defence, and the building and remodelling of exoskeleton chitin. In other words, arthropods have the ability to adapt quickly to new ecosystems by dramatically altering their shape and structure; they can cope with chemically challenging environments; and they have been able to evolve to eat just about anything!

Crustaceans are an exceptionally successful animal group by any standard. Their almost 67,000 described species may seem small compared to the vast number of insect species, but crustaceans are as abundant in the oceans as insects are on land. It has been estimated that tiny marine copepod crustaceans make up more than half of all animals in the world by sheer numbers, and krill (the favourite food of some whale species) has one of the greatest biomasses on the planet. As such, crustaceans are a crucial component of most marine food webs.

Unlike other arthropods, crustaceans have two pairs of antennae, although one pair can sometimes be tiny and difficult to see. Crustaceans are also unique in sharing a special first larval form known as a *nauplius* (though sometimes some or all larval stages can be very short and even take place within the egg before release). Most large crustaceans have well-developed gills, but some smaller species respire directly through the body wall. Like other arthropods, crustaceans have a stiff exoskeleton, which must be shed regularly to allow the animal to grow.

Crabs belong to the crustacean order Decapoda (literally meaning 'ten legs'), which also includes shrimps, prawns and lobsters – in fact, almost all of the well-known, edible and commercially important species. The ten legs are arranged in five pairs, with the first pair, or sometimes the second, often being enlarged and modified as claws for feeding, and for attack and defence.

In popular parlance, 'crab' is used for a number of different crab-like arthropods that are actually not closely related. True crabs belong strictly to a subgrouping of decapods known as the infraorder Brachyura. So, who are these impostors?

ABOVE: The giant Japanese spider crab (*Macrocheira kaempferi*) is the largest known arthropod – it can reach 3.8 m (12 ft) in claw span, 40 cm (16 in) in carapace width, and 19 kg (42 lb) in weight.

HORSESHOE CRABS

Horseshoe crabs have a passing resemblance to some crabs, but are not even crustaceans! Despite being tied to the sea, their body plan is quite different, and they are actually arachnids, in the order Xiphosura. New genetic evidence suggests they may have shared a common ancestor with 'hooded tickspiders' (order Ricinulei), another small and ancient group of arachnids. The four species of horseshoe crab that survive today all belong to a single family, the Limulidae. They are considered 'living fossils', looking much like their ancestors, which first appeared in the oceans around 450 mya.

Horseshoe crabs typically live in shallow coastal waters on soft sandy or muddy bottoms, but tend to spawn on intertidal mud flats or sandy beaches. In North America, the large spring and summer nesting aggregations of the Atlantic horseshoe crab (*Limulus polyphemus*) on the beaches of Delaware, New Jersey and Maryland are world-famous. Unfortunately, their future looks somewhat bleak as a consequence of coastal habitat destruction, and from overharvesting – both for food and for their blood, which is used in the biomedical industry to test for contamination.

LEFT: Atlantic horseshoe crabs come ashore on spring high tides to lay their eggs. Females make a nest in the sand at a depth of 15–20 cm (6–8 in) in which they lay as many as 64,000 eggs.

THE ANOMURANS

The rest of the 'would-be' crabs belong to the infraorder Anomura – a diverse group that includes numerous families, some of which are crab-like, and some more lobster-like. 'Anomura' literally means 'differently tailed' and refers to their variety of unusual pleons (abdomen or tail). Particularly strange are the hermit crabs that protect their typically soft, bloated tails by hiding them in discarded snail shells (see page 28). However the pleons of most anomurans are relatively reduced in size, and flattened, compared to those of shrimps and lobsters. In some, the pleon is carried as a broad flap tucked under the body, making them look very crab-like – an evolutionary pathway referred to as 'carcinization'.

Mole crabs and sand crabs

Hippoid 'crabs' (superfamily Hippoidea) mainly fall into the families Hippidae and Albuneidae. Mole and sand crabs are specialists at burrowing in sand, often in the surf zone of sandy beaches, so they have evolved remarkably similar body shapes to the raninid frog crabs that also inhabit sandy environments. Hippoid crabs have not lost typical anomuran features, especially a broad abdomen with a tail fan for swimming and escape propulsion, although unlike many anomurans, the tail is largely tucked beneath the body. Species in the family Hippidae can also be immediately distinguished from true crabs because they lack claws on the first pair of legs.

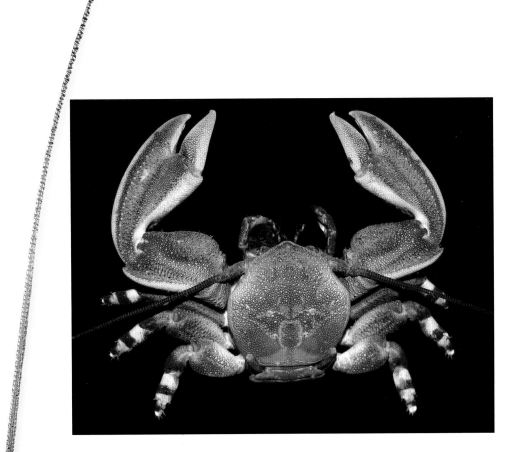

OPPOSITE: A mole crab (*Emerita analoga*). Not a true crab but an anomuran, this small crustacean burrows in the sand in the surf wash zone and uses its antennae for filter feeding.

RIGHT: The violet porcelain crab (*Petrolisthes violaceus*) is a southeastern Pacific species that inhabits intertidal rocky shores along the coast of Chile.

BELOW: The priest sand crab (*Albunea symmysta*) is rarely seen because it completely buries itself. Often the only sign of its presence is a small hole in the surface from which its thin, remarkably long antennae may protrude.

Porcelain crabs

This family of anomurans superficially resembles true crabs by having large flattened claws, rounded bodies, and the lack of a visible tail from above. Their true nature is revealed, however, in a flattened pleon that retains the swimmerets (see page 71) in both males and females, and ends in a typical lobster-like tail fan. This shows them to be much closer to squat lobsters than brachyuran crabs. Besides the claws, they also possess only three obvious pairs of walking legs, the last pair being much reduced, held up against the carapace, and mainly used for grooming. Around 280 species have been described. They have a cosmopolitan distribution in tropical and temperate waters around the world. Unlike almost all true crabs, they are suspension feeders and use long feathery mouthparts to comb the water for plankton and suspended organic matter.

Hermit crabs

There are six modern families of hermit crab, which include over 1,100 species. In fact a 'naked' hermit really looks very little like a crab because they typically possess a soft, swollen and asymmetrical tail. The reason they have been given the crab epithet is because the tail is characteristically hidden inside a snail shell, and all that is visible are a large pair of crab-like claws, some legs, and a pair of eyes poking out from the shelter of its mobile home. Most hermits are marine, but one family, Coenobitidae, contains a group of semi-terrestrial tropical species that live behind the beach-line and even into coastal forests. Most of these are small and still use snail shells like their relatives in the sea, but one, the robber crab (*Birgus latro*; see page 44), grows very large, and has an armour-plated, shortened tail, enabling it to dispense with the need for shells altogether.

Hairy stone crab

The hairy stone crab (*Lomis hirta*) is a strange little crustacean – only about 2.5 cm (1 in) wide, and the only representative of its own family, and superfamily (Lomisoidea), within the Anomura. A slow-moving, inoffensive animal, its compact shape and covering of short brown hairs give it excellent camouflage amongst the intertidal rocks along the southern Australian coastline.

King crabs

King crabs (Lithodidae) – also known as stone crabs – are the most crab-like of all anomurans, and form the basis of some very important commercial fisheries in the northern hemisphere. They were long thought to be a separate group, but genetic analyses have now shown that they evolved relatively recently from pagurid hermit crabs, in much the same way that the king hermit crab (*Patagurus rex*) did. The tail is much reduced in size, flattened and completely tucked up under the body, much as it is in true crabs, but it is also asymmetrical, like the tail of its hermit crab-like ancestors. King crabs are particularly good examples of carcinization because they look and act so much like real crabs (see also page 202).

OPPOSITE: The pink-clawed hermit crab (*Dardanus pedunculatus*) is an Indo-West Pacific species that inhabits subtidal reefs and seagrass beds. Like several hermits, it plays host to sea anemones.

ABOVE: The king hermit crab (*Patagurus rex*) was found at 400 m (1,300 ft) in French Polynesia. With its tiny tail and broad, flattened carapace, it is the most crab-like of any hermit. It carries a small, flat oyster shell to protect its tail, but its long legs mean it can run quickly from predators.

LEFT: The hairy stone crab is the only representative of a group that diverged from other anomurans around 120 mya.

CRABS IN PREHISTORY

THE FOSSIL RECORD IS AN ESSENTIAL TOOL for understanding the evolution of life on our planet, but it is imperfect. For a living organism to be captured for eternity in stone is a fortuitous event, but to be rediscovered many millions of years later, and for its story to be told, is even more remarkable. Nevertheless, there are now over 3,000 known fossil species of Brachyura, with what is arguably the earliest, *Eoprosopon klugi*, dating from the Lower Jurassic Period around 185 mya. (Another, *Eocacinus praecursor*, appeared immediately after the Triassic-Jurassic extinction event, around 200 mya, but its position as a brachyuran is still under question.) *Eoprosopon klugi* is placed in the family Homolodromiidae, which includes crabs looking much the same today – truly living fossils.

THE MESOZOIC MARINE REVOLUTION

To properly understand the origins of crabs it is necessary to go further back in time – over half a billion years ago. The Palaeozoic Era, starting with the Cambrian Period of 570 mya, gave rise to a fantastic explosion of life in the sea. The very first crustaceans appeared at this time, but the seas were dominated by another extremely diverse arthropod order, the trilobites, which swarmed over the seabed as predators, scavengers and filter feeders. Crustaceans also shared those primeval seas with brachiopods (lamp shells), sponges, shelled molluscs, worms, jellyfish, early echinoderms, and eventually nautiloids and early corals. This first great blooming of life ended abruptly 252 mya, at the end of the Permian, with the greatest extinction in

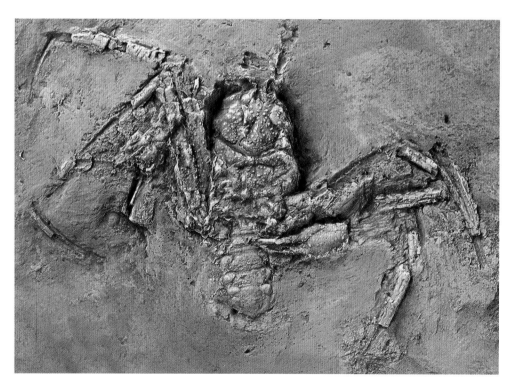

LEFT: *Eoprosopon klugi* – the oldest fossil incontrovertible brachyuran crab. Discovered in southern Germany, it lived around 185 mya. It is considered part of the family Homolodromiidae, which has living representatives to this day.

OPPOSITE: *Callichimaera perplexa*, from 90–95 mya, evolved during a period of morphological experimentation in the mid-Cretaceous, and is believed to represent the first marine arthropod to evolve highly modified, flattened oar-like legs for swimming.

Earth's history, the 'Great Dying'. Large-scale volcanic activity in today's Siberian region led to global warming. As sea temperatures rose, the metabolism of marine animals sped up, and the warmer waters could not hold enough oxygen for them to survive – perhaps 96 per cent of all species were lost.

Thus, the beginning of the Mesozoic Era (250–66 mya) was marked by a major change in the ecology of the sea. The reset button had been pushed, and life had essentially to start again from the remnant survivors of the past. The widespread evolutionary changes amongst benthic invertebrates during this time are referred to as the Mesozoic Marine Revolution.

At its beginning, during the early Triassic, the world's oceans were still characterized by very low oxygen levels, elevated carbon dioxide, and ocean chemistry that reduced carbonate deposition, which made it very difficult for most animals to absorb calcium. However, modern stony corals suddenly appear in the fossil record in the mid-Triassic Period (around 230 mya), indicating that oxygen levels and water chemistry had returned to a state whereby marine animals could accrete calcium from seawater and begin to build skeletons efficiently again. Of course, a strong, calcium-fortified exoskeleton was crucial for crabs, and a necessity for their evolution and diversification.

While the first fossil crab dates to only a little later than corals, at around 200–180 mya, it seems likely that they were present somewhat earlier. Unfortunately crabs and other decapod crustaceans seem not to have been well preserved as fossils through the Triassic. The earliest anomuran (*Platykotta akaina*) appears around the late Norian to Rhaetian Ages of the Late Triassic (227–201 mya), and as anomurans and brachyurans evolved from a common ancestor, this strongly suggests that crabs may be of similar antiquity. Because trilobites had already become extinct at the end of the Permian, the seabed was now available to be ruled by a new form of dominant arthropod – the crab!

THE LATE JURASSIC PERIOD

The beginnings of what was to be the first evolutionary explosion within Brachyura occurred towards the end of the Jurassic (about 150 mya), with dozens of species and genera appearing in the fossil record. It was also the time that the supercontinent Pangaea split into Laurasia to the north and Gondwana to the south; as these landmasses also began to split apart, lengthy new coastlines and large new shallow seas formed the perfect environments for nascent forms to diversify. The earliest known fossils were represented predominantly by primitive podotreme crabs. Many of these lived on or in more open substrates, and developed special limb modifications to carry objects to conceal themselves from predators. Also during the Late Jurassic, planktonic microbiota began to flourish in the upper sunlit seas, both in quantity and diversity, so there were now large phytoplankton blooms photosynthesizing and sequestering carbon and energy from the sun. The phytoplankton was in turn eaten by zooplankton, and as all these floating organisms died, they settled to the bottom, providing the energy to drive benthic ecosystems and more complex food chains. Crabs became mobile scavengers and detritovores to take advantage of this abundant new food source.

Throughout the Jurassic, the growth of coral reefs, with their complex structures and biological complexity, also gave crabs many new niches in which to expand. By the end of the Jurassic diversification, ten superfamilies, and many more families of Podotremata and Archaeobrachyura, had evolved. However, life in the oceans took another big hit when a massive asteroid struck Mexico's Yucatán Peninsula. The resultant catastrophic Cretaceous–Paleocene extinction event of 66 mya caused the end of the dinosaurs, and it also had a serious, but less devastating, impact on marine invertebrates. It marked the end of the period of dominance of the 'primitive' crabs, allowing another group to flourish and diversify: the Eubrachyura, or true crabs, that still dominate today.

BELOW: *Periacanthus horridus* lived during the Middle Eocene, about 48 mya. This specimen was found in Vicenza, Italy. It is an early spider crab; its closest relatives today would be in the family Epialtidae.

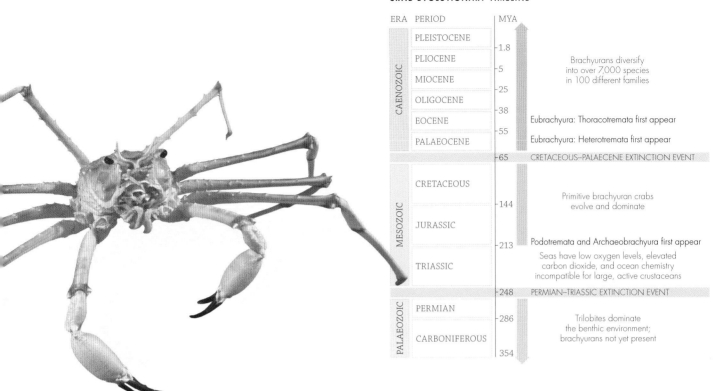

CRAB EVOLUTIONARY TIMELINE

ERA	PERIOD	MYA	
CAENOZOIC	PLEISTOCENE	1.8	
	PLIOCENE	5	Brachyurans diversify into over 7,000 species in 100 different families
	MIOCENE	25	
	OLIGOCENE	38	
	EOCENE	55	Eubrachyura: Thoracotremata first appear
	PALAEOCENE	65	Eubrachyura: Heterotremata first appear
			CRETACEOUS–PALAEOCENE EXTINCTION EVENT
MESOZOIC	CRETACEOUS	144	Primitive brachyuran crabs evolve and dominate
	JURASSIC	213	Podotremata and Archaeobrachyura first appear
	TRIASSIC	248	Seas have low oxygen levels, elevated carbon dioxide, and ocean chemistry incompatible for large, active crustaceans
			PERMIAN–TRIASSIC EXTINCTION EVENT
PALAEOZOIC	PERMIAN	286	Trilobites dominate the benthic environment; brachyurans not yet present
	CARBONIFEROUS	354	

ABOVE: An antlered crab (*Dagnaudus petterdi*) belongs to the Podotremata, an ancient group of crabs that first appeared about 150 mya. Like their ancestors they live on open substrates, and typically carry objects on their backs to hide under, and protect them from predators.

Modern fishes, which had become widespread towards the end of the Cretaceous, also become abundant during the Palaeocene. These highly mobile predators would have exerted much pressure on crabs to quickly evolve defensive strategies, including an agility to escape quickly into hiding, camouflaged body shapes, and decorating behaviour. Finally, the Thoracotremata, the most recently evolved group of crabs, and the group that has conquered the challenging intertidal and terrestrial environments, began to appear during the Lower Palaeocene, and underwent rapid diversification into numerous families during the Eocene (55–38 mya). Thus, the basis of the crab fauna we know today was born.

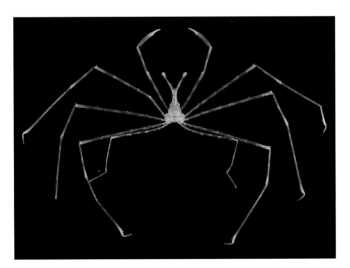

LEFT: The stalk-eyed spider crab (*Latreillia metanesa*) has a fossil record dating from the Upper Cretaceous. Long, spindly legs allow it to move quickly across sand, coral and rock at depths of 300 m (985 ft).

CRABS IN PREHISTORY

EVOLUTIONARY TRENDS WITHIN BRACHYURANS

THE FRENCH ZOOLOGIST PIERRE LATREILLE (1762–1833) worked on arthropods at the Muséum National d'Histoire Naturelle in Paris, and was considered the foremost entomologist of his time. In 1812, he made the first serious attempt to group crabs into some 'natural order'. Latreille recognized seven 'sections' based on the overall body shape and arrangement of the legs: *Nageurs* (swimming crabs with flattened legs), *Arqués* (crabs with arched or arcuate carapaces with pointed dactyli), *Quadrilatères* (square to rectangular-shaped crabs), *Orbiculaires* (rounded or elliptical crabs), *Triangulaires* (triangular or rhomboidal crabs), *Cryptopodes* (crabs with legs hidden by the carapace) and *Notopodes* (last two pairs of legs placed anteriorly on carapace and forming claws).

This first classification was based only on superficial similarities, and these seven sections have now been divided into 100 families in four major divisions, although two sections still reflect major groupings we recognize today: *Nageurs* mirrors today's family Portunidae, and *Notopodes* includes all the 'primitive' crabs that we now place in the sections Podotremata and Archaeobrachyura.

As more and more crabs were found and described, Latreille's classification was replaced by increasingly sophisticated proposals that attempted to look at evolutionary patterns beyond the superficial shapes of the carapaces. Many new family groups of crabs continued to be proposed and described, and there was divided opinion on how best to group them, so it was only in the 1960s that the basic principles of our modern 'higher order' understanding of crab evolution were finally proposed. Essentially, this went back to basics by trying to understand the process of carcinization, and how different lineages had improved on being a crab in the modern world. This meant examining the internal structures of the skeleton, and, in particular,

ABOVE: Examples of 'Nageurs' figured in *The Animal Kingdom, Arranged According to its Organization…* (1834). Pierre Latreille's work provided a foundation for the classification of crabs for many years.

changes in the physiology of reproduction. It was proposed that there is a large group of 'primitive' crabs that comprises a separate major lineage (this was named Podotremata), and a second much larger sister group that includes the greatest number of the modern crabs living today, the Eubrachyura.

Podotremes are most similar in appearance to their anomuran ancestors, partly because they are usually longer than wide, and partly because the male pleon is usually broader and not locked into place under the body; but more importantly, these crabs all still have both female and male genital openings positioned on the coxa of the walking legs (the second and the last respectively, see page 63) – the name Podotremata is derived from ancient Greek *podo* (meaning 'foot') and *trema* ('hole'). This pattern provided strong evidence of their link to anomurans such as hermit crabs, porcelain crabs and king crabs. However, the latest genetic research has shown that the Podotremata actually consists of two sister groups that have been long separated from a more primitive ancestor again. This has meant acknowledging two groups of primitive crabs, one keeping the name Podotremata, and the other being given the name Archaeobrachyura.

Importantly, both podotremes and archaeobrachyurans retain external fertilization of their eggs, unlike the Eubrachyura, in which the female opening has migrated from the legs to the middle of the sternum (see page 72). There are two distinct groups of eubrachyurans – the Heterotremata and the Thoracotremata (see pages 18–19), each separated according to the position of the male gonopore and penis. In heterotremes, the male gonopore still originates from the coxal segment of the fifth leg, whereas in thoracotremes, the male openings open directly through pores on the base of the sternum beneath the pleon. This allows the reproductive organs to act independently of the legs, and seems to have been particularly important in allowing crabs to conquer intertidal and terrestrial environments, as these habitats are almost the sole province of the thoracotreme families.

LEFT: The deepwater orange velvet crab (*Metadynomene tanensis*), belongs to the Dynomenidae, another ancient family of podotreme crabs. Strangely, the much-reduced last pair of legs of crabs in this family seem to have no discernible use.

EVOLUTIONARY TRENDS WITHIN BRACHYURANS

CONVERGENT EVOLUTION

Convergent evolution is the name of the process that leads unrelated species to evolve similar appearances and habits, typically because they are adapting to the same type of environment, or ecological niche. It is normal for humans to group things that look similar, and this is indeed what the early naturalists did. For example, barnacles were classified as molluscs for a long time because their hard shell glued to rocks make them *look* like molluscs. But on closer examination, the animal hiding inside the shell was found to be a small crustacean lying on its back and poking its feathery legs out to feed. This emphasizes the fact that an evolutionary biologist must not jump to conclusions based on superficial similarities. Scientists strive to classify into 'natural' groups, which means that they can recognize direct ancestry (*monophyly*). When it is discovered that two species have been grouped together, but have had different ancestors (thus are not really closely related), then that group is said to be *polyphyletic*, and the species need to be separated into different genera, or families, that include their real relatives. Recognizing polyphyly is sometimes a case of looking harder, but increasingly genetic DNA evidence is being used as a powerful tool to show real interrelationships.

BELOW: The guard crabs are a great example of convergent evolution for living a life in coral. Long thought to all belong to a single family because of their remarkable similarity in appearance, it was only in 2004 that it was realized that two separate families of crabs had evolved independently of each other. The crab on the left, *Tetraloides heterodactyla*, belongs to the Tetraliidae, while *Trapezia cymodoce* on the right is in the family Trapeziidae.

RIGHT: Elongate eyestalks have evolved independently in a variety of different types of crab. This is an adaptation to living in burrows on flat, open substrates, and it enables them to see mobile predators with plenty of warning. The stalked eyes of this subtidal petal-eyed swimming crab (*Ommatocarcinus minabensis*) are much like those of intertidal fiddler and sentinel crabs.

A fascinating genetic study on Jamaican freshwater and semi-terrestrial crabs has provided novel insights into how evolution works, and how quickly it can happen. It illustrates clearly how morphological divergence (changes in form and structure) is caused, and driven by, ecological convergence.

Geological evidence shows that the island of Jamaica only emerged from the Caribbean Sea during the early Pliocene, so the complex ecosystems of today, with their rich diversity, only developed over the last 4 or 5 million years. The unoccupied terrestrial and freshwater habitats of this new landmass proved ripe for exploitation by emigrant intertidal crabs. Today there are ten endemic freshwater and forest crabs that all belong to the family Sesarmidae. Elsewhere in the Americas, sesarmids are typically coastal mangrove and estuarine inhabitants, but in Jamaica six species have become completely freshwater in habit, while the other four thrive in caves and more terrestrial habitats. It was assumed that the island must had been colonized by several different ancestral crab types over time, because many of the crabs look so different, and have such unique physiological and behavioural adaptations (such as active brood-care for larvae and juveniles). In support of this idea, one species was included in the Southeast Asian genus *Sesarmoides*, while another was considered so different that it was placed into its own endemic Jamaican genus, *Metopaulius*. Amazingly, though, the study showed that all Jamaican land crabs originated from a single marine sesarmid ancestor that only invaded the island within the last 4 million years. By contrast, marine species separated for about the same amount of time by the emergence of the Isthmus of Panama between North and South America, forming a barrier between the Pacific and Atlantic Oceans, are still almost identical. In the case of Jamaica, the sudden opportunity to invade new habitats and niches caused rapid evolution into multiple new forms.

SPECIATION AND CRYPTIC DIVERSITY

Because the oceans of the world are largely connected, it was long assumed that most marine species were able to journey great distances without obvious barriers. However, the sea is not an homogeneous environment – the oceans, and often the seas within them, have internal circulation patterns and temperature gradients that tend to keep larval dispersal confined.

Smaller-scale current patterns can keep larvae restricted and separated from even nearby localities, and the reproduction strategies of individual species also play an important role in how quickly larvae develop, and how exposed they are to the open ocean currents. Many marine crabs have long-lived larvae, giving them plenty of time to travel widely, and thus maintain genetic connections over large distances, while others can be restricted to a short stretch of coastline.

BELOW: Crabs, along with most marine animals, have evolved independently in different regions, with little overlap in species composition. These regions are separated by ocean current circulation patterns, winds, temperatures and depths – factors that combine to keep larval dispersal confined.

MAJOR MARINE BIOGEOGRAPHIC REGIONS

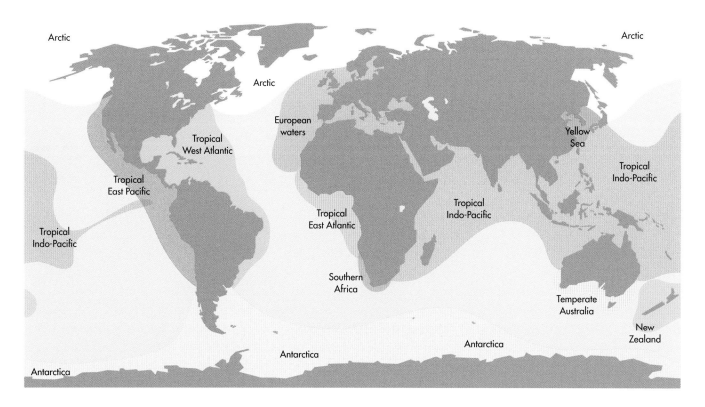

In almost all cases, speciation is driven by barriers to gene flow. If two populations of a species are separated by some geological event ('vicariance') long enough, they begin to evolve and change until eventually they can no longer interbreed. This is termed 'allopatric speciation', and can occur quickly or slowly; for example, when the ancient break-up of the continents formed new oceans, or as recently as the last few hundred thousand years, when fluctuating sea levels created land barriers that isolated the crabs on either side. Speciation is a dynamic and continuous process.

With the extensive use of genetic techniques in recent years it has become apparent that there are many unrecognized and undescribed species that visibly differ from each other in only small ways. Of course, just as in humans, natural variation occurs between individual crabs, but the trick is in understanding when such differences become large and consistent enough to represent populations of separate species.

BELOW: Most crabs disperse by casting their larvae into the currents, but adults of the genus *Planes* live a nomadic life afloat on the oceans as members of diverse neustonic communities. Clinging to mats of algae, logs, fishing floats and even on the shells of sea tutrles, they are mostly only encountered when their home is washed ashore by onshore winds.

Increasingly, what were once thought to be widespread species are now proving to be complexes of separate species. The Indo-West Pacific blue swimmer crab is a good example. It was assumed to be a single species, *Portunus pelagicus*, that had spread from East Africa to Japan, Australia and east into the Pacific. However, an intensive study looking at samples from across its range showed that four species were actually mixed up together. Genetic analyses showed clear separations, and this allowed taxonomists to get a good understanding of how much variation in colour and shape occurred within each of the species, which led to the development of identification tools that could be used by fisheries agencies to properly manage them individually.

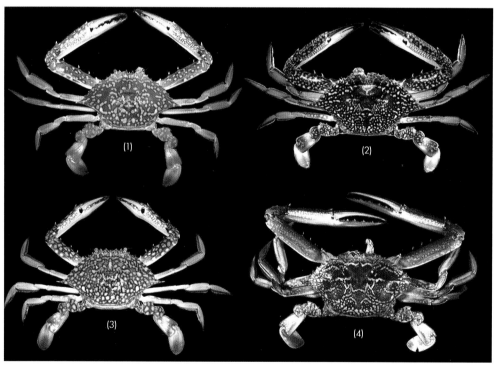

ABOVE: The purple-ocellated reef crab (*Xanthias maculatus*) is found in coral reef environments. There are several similarly spotted species through the Indo-West Pacific, suggesting that new species are currently being confused under this name. Genetic studies may help to separate them.

LEFT: An example of careful study revealing previously confused cryptic speciation – for many years, four separate species were considered to be simply geographic variations of the widespread Indo-West Pacific blue swimmer: 1. *Portunus pelagicus*, 2. *Portunus segnis*, 3. *Portunus reticulatus*, 4. *Portunus armatus*.

The analyses also showed which of the species were more closely related to each other, and approximately how long ago each diverged from its sister species – the ancestral *Portunus pelagicus* had first split into two species around 5.5 mya, and then again at 3.3 mya, again at 2.5 mya, and finally about 800,000 years ago. Studies on the Indo-West Pacific giant mud crab (*Scylla serrata*) have revealed an almost identical story, and resulted in it also being split into four sibling species.

Understanding the events that lead to speciation in the sea can be quite difficult, and relies heavily on our increasing knowledge of palaeogeography – particularly how continents have drifted, how climate and ocean circulation patterns have changed, and how changes in sea levels have affected the formation of land bridges. Sometimes events leading to speciation are relatively straightforward – one well-known example being the formation of the Isthmus of Panama, mentioned earlier – but overall there is still much to learn.

BELOW: Studies of the commercially important Indo-West Pacific giant mud crab revealed that four different species had been confused together. Recognition of these differences has been of great importance in improving aquaculture techniques, and for managing wild fisheries.

The evolution of freshwater crabs reveals a much simpler and clearer picture. Because these crabs give birth to live young, their ability to disperse is largely dependent on adult migration. The adults are dependent on fresh water, however, and do not like to stray too far from it. In hotter, wetter times of the past, ancestral crabs were able to walk from river to river, and gradually spread across Asia and Africa, and even south into Australia. But as climates have become more arid, and river catchments have become isolated, they have become constrained in their movements, and gradually begun to evolve into separate species (despite often still looking very similar). Ongoing studies of the Australian crab genus *Austrothelphusa* are revealing that there were several 'invasions' of Australia in the past, as crabs moved down from New Guinea across Torres Strait land bridges during wet periods, then spread out to become the progenitors of the numerous species seen today. Genetic studies are

showing that almost every separate river catchment around the coast has its own species, most of which are still formally undescribed; each took refuge in its river system after northern Australia began to dry out and lose its tropical forests, around 2 mya. Similar stories abound from wherever freshwater crabs occur around the world.

ABOVE: A triangle crab (*Cryptopodia dorsalis*) has a broad, flattened body that looks just like broken coral or shell.

LEFT: *Arachnothelphusa terrapes* lives exclusively in water-filled holes in trees ('phytotelmata'). It is also nocturnal, extremely sensitive to light, and only found in the rainforests of Sarawak, Indonesia. Little wonder it is rarely seen.

OPPOSITE: While appearing like colourful little gems on the intertidal mangrove mud, these male fiddler crabs (*Tubuca coarctata*) stage fierce territorial battles to protect their burrows from would-be usurpers.

ROBBER CRAB
Birgus latro

FAMILY:	Coenobitidae
OTHER NAMES:	Coconut crab
DISTRIBUTION:	Widespread on oceanic islands through the tropical Indo-West Pacific region
HABITAT:	Terrestrial
FEEDING HABITS:	Omnivore, predator and scavenger
NOTES:	Extremely slow growing but can live for 40–50 years. Prized as food, populations have been severely depleted on many islands.
SIZE:	Over 4 kg (9 lb) in weight; 50 cm (20 in) in length

THE ROBBER CRAB is the world's largest land-living invertebrate and can be an intimidating animal to meet in the forest. Despite its name it is not really a 'true' crab but closely related to hermit crabs, though unlike hermits, it no longer needs to use a shell to protect its pleon. While it must go to the sea to release its eggs, it is otherwise fully terrestrial, and will drown if submerged in water. Unfortunately, in only 50 years *Birgus latro* has declined from prolific across its range to being listed as 'endangered' by the International Union for the Conservation of Nature. Over-exploitation has led to extinction of some island populations in less than a decade.

VORACIOUS FEEDERS

These crabs will eat virtually anything they can catch, and are voracious scavengers. It has even been suggested that the lost aviatrix Amelia Earhart, who crashed in Kiribati in the south Pacific Ocean, may have been consumed by the abundant populations of robber crabs that live on those islands. To test this theory, a pig carcass placed on the ground attracted many robber crabs as well as swarms of the smaller strawberry hermit crabs (*Coenobita perlatus*); the pig was consumed within two weeks, with some bones being dragged as far as 20 m (66 ft) away! Although slow-moving, robber crabs are strong, agile and excellent climbers, capable of scaling rock faces and tall palm trees in search of fruits. Coconuts are a preferred food, and they can peel and crack the fresh nuts with their immensely powerful claws.

A COCONUT OBSESSION

'Robbers' will go to great lengths to satisfy their coconut cravings – even climbing tall trees to get the fresh fruit. Unfortunately, its formidable claws are not enough to protect it from human predators, and it is now endangered in large parts of its range.

HALLOWEEN HERMIT CRAB
Ciliopagurus tricolor

HERMIT CRABS ARE FAMOUS for hiding their bodies away so that all that can be seen appears to be a normal crab poking out its legs and claws from the confines of a marine snail shell. But, free of the shell, they show their true colours as an anomuran with a long, fleshy tail. There are over 1,100 species of hermit, grouped into around 120 genera and six families. Recent genetic studies reveal that the closely related king crabs (see page 48) evolved from hermit crab ancestors (in particular the family Paguridae). With the pleon shrunken and flattened under the body, king crabs no longer have to lug around heavy shelters, and this has enabled them to become much much larger, and highly successful in exploiting new habitats.

BORROWING A HOME

Halloween hermits, like most other species, typically possess a large, soft, swollen and asymmetrical tail that is twisted dextrally (to the right) to suit the coiling direction of gastropod shells, making it easy to slip in and out. The specially modified uropods of the tail fan have rasp-like raised surfaces that can be pushed against the inside of the shell to hold the hermit tightly in place. The snail shell is used very much like a second exoskeleton, so when hermits grow and need to moult, they must also find a bigger shell to wear. They can be quite picky when looking for the perfect fit, and may try on several for size before making their choice.

FAMILY:	Diogenidae
OTHER NAMES:	Cone-shell hermit crab, orange-legged hermit crab, striped hermit crab
DISTRIBUTION:	Eastern Africa south of Somalia; Madagascar, Mayotte and Reunion in the southwestern Indian Ocean. Similar-looking species of *Ciliopagurus* are found throughout much of the Indo-West Pacific region.
HABITAT:	Sandy areas around coral reefs; low intertidal to shallow subtidal to about 15 m (50 ft) depth
FEEDING HABITS:	Omnivores and scavengers; diet includes algae, cyanobacteria, detritus, and whatever it can find
NOTES:	*Ciliopagurus* species are all flattened and perfectly adapted to the narrow opening of cone shells. Can live 8–10 years in aquariums.
SIZE:	Up to about 5 cm (2 in) in length

CLOSET FASHIONS

Unlike many hermit crabs, the halloween hermit continues its flamboyant colouring over its whole body, even though the pleon is normally hidden deep inside a snail shell.

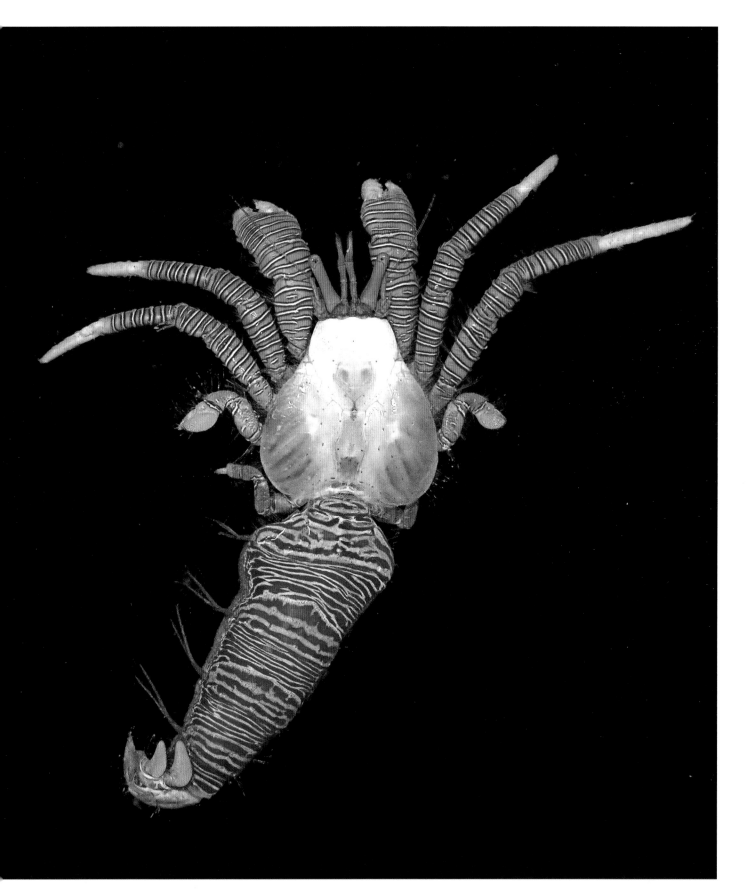

PUGET SOUND KING CRAB
Lopholithodes mandtii

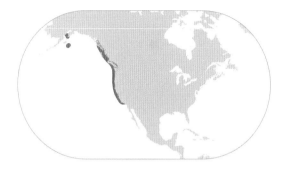

FAMILY:	Lithodidae
DISTRIBUTION:	US Pacific Coast, from Sitka (Alaska) to central California
HABITAT:	Shallow subtidal to about 140 m (460 ft); prefers rocky areas
FEEDING HABITS:	Generalist invertebrate predator, especially of echinoderms such as sea urchins, starfish and sea cucumbers, and also barnacles and anemones
NOTES:	Now relatively rare; on the Washington Department of Fish and Wildlife's protected species list
SIZE:	Up to about 30 cm (12 in) carapace width

ONE OF THE KING CRABS in the family Lithodidae, the Puget Sound king crab is not a true crab, but belongs to the infraorder Anomura along with a number of other lookalikes. King crabs are the most crab-like of all anomurans with their tail much reduced in size and completely tucked up under the body; but unlike proper crabs they only have three pairs of legs visible, and still retain a tail fan and an asymmetrical pleon like their hermit crab-like ancestors.

MAKING LIKE A ROCK

This is the largest 'crab' (either brachyuran or anomuran) on the West Coast of the United States, at least in carapace width if not in leg span – that honour goes to another lithodid, the red king crab (*Paralithodes camtschaticus*). Adults prefer rocky areas in deeper water, often in areas of strong current flow. Their strong, pointed claws can dig into crevices in vertical walls, and they can perch on small ledges. Juveniles prefer to be under rocks during low tide. When threatened, it tightly locks its stout legs and strong claws against its heavy shell, forming itself into an impenetrable fortress against would-be predators.

A MASTER OF DISGUISE DESPITE ITS SIZE
The rough, uneven texture of the shell, with its bumps and hollows, matches the crab's rocky environment, and its mottled colours complement those of the surrounding algae and other encrusting organisms.

DURIAN CRAB
Acanthodromia species

DYNOMENIDS ARE A SMALL GROUP of 'primitive' crabs living in tropical and subtropical waters of the Atlantic, Indian and Pacific Oceans. Their closest relatives are believed to be the dromiid sponge crabs. They are immediately recognizable by their broad, triangular frontal margins. Most dwell in shallow reefs, while those living in water deeper than about 100 m (330 ft), seem to live on lithothamnion algae, red coral and precious coral, as well as rock and sand. The depth record for the family is held by *Acanthodromia erinacea*, at 540 m (1,770 ft). Very little is known about the biology and ecology of dynomenids, but it seems they feed primarily by sieving organic fragments from the substrate, or by consuming coral mucus.

PARALLEL DEVELOPMENT

Acanthodromia would seem to have originated in the ancient Tethys Sea and then, as the continents drifted apart, one species became isolated in the western Atlantic and Caribbean, and another in the east Indian and west Pacific region. So the ancestors of the two species known today were most likely separated at least 65 mya, during the Palaeocene, as by this time the Atlantic was already well formed and the Caribbean isolated. Given their long separation it is truly remarkable that *A. erinacea* (Atlantic) and *A. margarita* (Indo-West Pacific) still look almost identical.

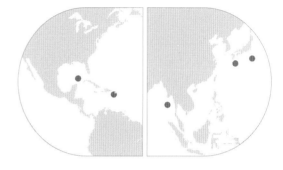

FAMILY:	Dynomenidae
OTHER NAMES:	Prickly dynomenid crab
DISTRIBUTION:	Bay of Bengal to southern Japan; central West Atlantic
HABITAT:	Not well known; probably reef patches on soft bottoms; 120–540 m (400–1,770 ft)
FEEDING HABITS:	Sieves organic matter from the substrate
NOTES:	Only 2 species described
SIZE:	To about 18 mm (¾ in) carapace length

A 'PRICKLY' NAME

Pictured here is *Acanthodromia margarita*, trawled from nearly 300 m (980 ft) off the Philippines. These crabs are so rarely encountered that no common name has ever been bestowed upon them. It is hereby christened the 'durian crab' because its prickly appearance so much resembles that tropical Asian fruit.

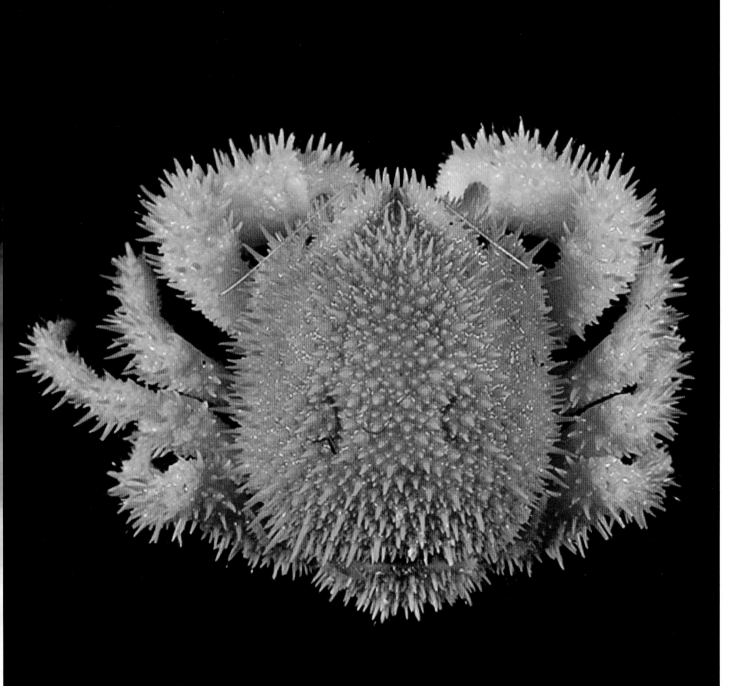

SLENDER FROG CRAB
Lysirude species

FAMILY:	Lyreididae
OTHER NAMES:	Spiny frog crab
DISTRIBUTION:	Tropical Indo-West Pacific; central western Atlantic (Massachusetts to Gulf of Mexico and Antilles)
HABITAT:	Sandy bottoms; deep sea
FEEDING HABITS:	Predator on benthic invertebrates
NOTES:	Distant relatives of the commercially fished spanner crab (*Ranina ranina*, see page 204)
SIZE:	To about 25 mm (1 in) carapace length

THESE ODD-LOOKING CRABS belong to the large superfamily Raninoidea, often termed 'frog crabs' because of their lengthened shape and the way they sit up from the bottom a bit like a frog. In fact, frog crabs are placed in the Archaeobrachyura, amongst the earliest and most primitive groups of crabs to have evolved. Similar raninoids first appeared in the Early Cretaceous around 145 mya, and are richly represented in the fossil record.

SAND SPECIALISTS

Frog crabs mostly live in soft, relatively sandy substrates, and can quickly bury themselves using their flattened legs, specifically evolved for efficient digging. They even have a specialized respiratory system that allows them to pump water through their gills while just their pointed snouts protrude from the bottom. Their sharp, pointed claws are deadly weapons against the worms, molluscs and other bottom-dwelling invertebrates that make up their diet. Most frog crabs live in relatively shallow coastal waters, but those in the small family Lyreididae (only six species in two genera) are restricted to the deep sea, from around 200 m to as much as 1,000 m (660–3,280 ft).

A NEW SPECIES?

This crab is most similar to *Lysirude channeri*, a species first described from the Bay of Bengal in 1885 by James Wood-Mason, director of Calcutta's Indian Museum. That species has since been recorded through to the South China Sea and the Philippines, but Pacific specimens differ in carapace spine number and length, so may prove to be one or more new species.

SHAGGY SHORE CRAB
Pilumnus vespertilio

PILUMNUS VESPERTILIO IS AN EXCELLENT EXAMPLE of how some crabs have evolved such strange appearances that they become almost invisible to predators. This moderately large crab is common on sheltered rocky shores and reefs throughout the tropical Indo-West Pacific, but surprisingly few shore-walkers would ever notice its presence. Crabs in the family Pilumnidae are commonly called 'hairy crabs', and there are none hairier than the shaggy shore crab. A thick growth of setae blurs the edges of their outline, providing extremely good camouflage. Found at low tide nestling into holes and rock crevices, usually with an added coating of silt, they are almost impossible to see unless they move. Being timid of predators, they are largely nocturnal feeders, emerging to graze towards twilight; even then, they are quick to scurry to shelter if disturbed. Their diet is primarily macro-algae of most types, even including coralline species, but their powerful claws are capable of dealing with a wide variety of intertidal animals, including sponges, zoanthids, polychaete bristle worms, brittle stars, marine snails, bivalves and sea slugs.

WHAT'S IN A NAME?
First described by the great Danish zoologist Johan Christian Fabricius in 1793, the derivation of this crab's name is a little unclear. *Vespa* is Latin for 'evening', while *tilia* refers to a bushy linden or lime tree, so perhaps the name describes the crab's nocturnal activity and thick covering. However, *Vespertilio* is also a genus of European bat, named 35 years earlier by Linnaeus, so perhaps it is the crabby namesake of a strange furry bat!

FAMILY:	Pilumnidae
OTHER NAMES:	Common hairy crab, bad-hair-day crab, mop crab
DISTRIBUTION:	Widespread in the Indo-West Pacific
HABITAT:	Rocky and coral rubble shores; low intertidal and shallow subtidal
FEEDING HABITS:	Omnivore; mainly algae but opportunistic predator of small encrusting and mobile invertebrates
NOTES:	Spawns year-round; egg release is timed to coincide with full moons, when higher tides will suck zoea safely away from the shore
SIZE:	To about 40 mm (1 5/8 in) carapace width

MOP HEAD
The shaggy covering over the top of the body and legs completely obscures the outline of the crab so when it is resting on the bottom it looks like a patch of algae. The thick setae are also a magnet for muddy silt, which further disguises its presence.

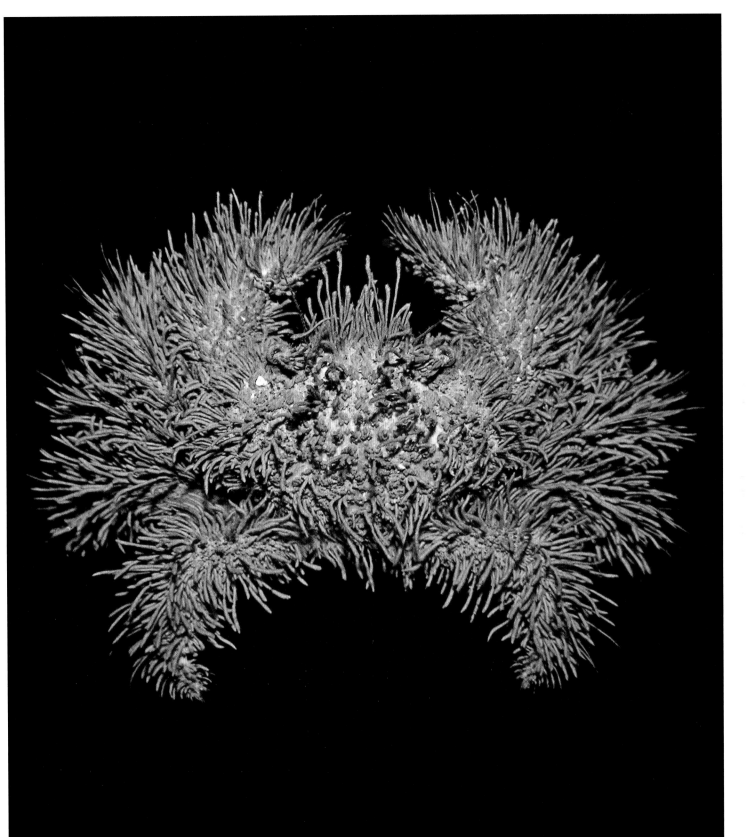

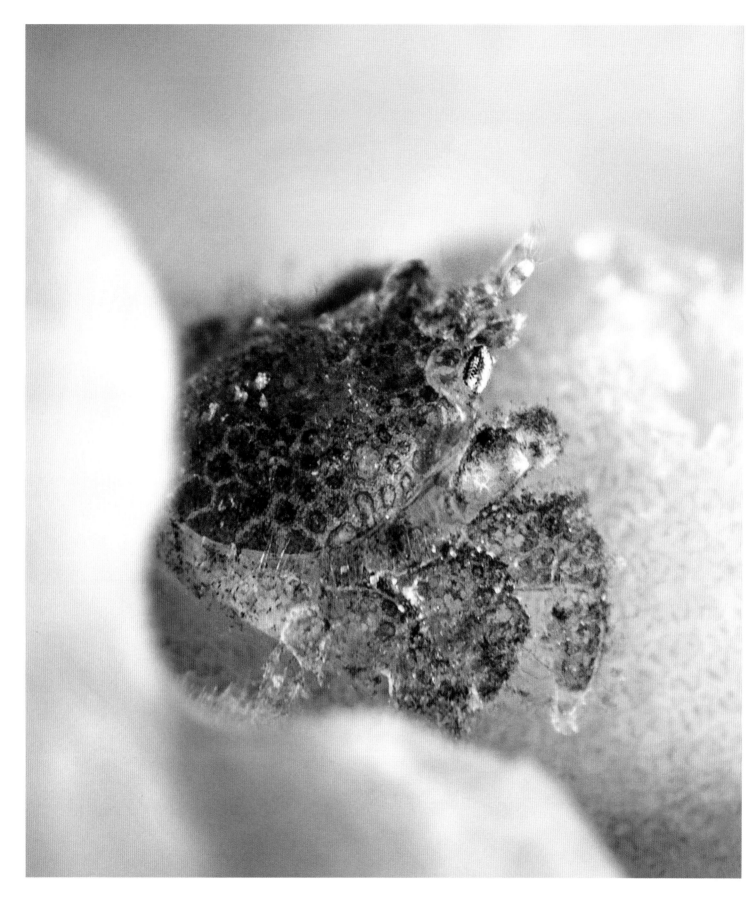

GREEN-SPOTTED GALL CRAB
Pseudocryptochirus viridis

FAMILY:	Cryptochiridae
DISTRIBUTION:	Central Indo-West Pacific; Japan south to Indonesia, Guam; Palau and Caroline Islands
HABITAT:	Shallow crescent-shaped pits on colonies of *Turbinaria* coral; 1–15 m (3–50 ft) depth
FEEDING HABITS:	Eats coral mucus
NOTES:	Males are typically much smaller than females
SIZE:	To about 8 mm (3/8 in) carapace length

GALL CRABS (FAMILY CRYPTOCHIRIDAE) ALWAYS LIVE in symbiotic relationships with living stony corals. They are found in tunnels or pits in the surface, or in swollen 'galls'. Most of the 50 or so species each chooses a single species of coral host (or at least a similar one within the same family). Occurring in all tropical oceans, most live in the sunlit shallows, but one species has been found on a deep-water coral at 512 m (1,680 ft) depth. Females of the first gall crab to be discovered, *Hapalocarcinus marsupialis*, are forever trapped inside galls formed at the ends of coral branches, and from this strange 'marsupium' their eggs develop, and the larvae are released into the sea.

A ONE-WAY RELATIONSHIP?

While each species varies a little in its mode of feeding, they appear to all survive primarily on a diet of sticky coral mucus, and the debris that it accumulates. Some crabs deliberately irritate the coral with their legs and claws so it produces more for them to eat; some make mucus balls that they roll over the coral surface to sweep up debris before consuming it all. It is hard to know if the coral benefits, but the crabs do not seem to do any serious harm either. Cryptochirids belong to the most recently evolved Thoracotremata group of crabs, but they are so different from most other thoracotremes that their evolutionary relationships have long been mysterious. Modern genetic methods have now revealed that gall crabs, and the similarly strange and symbiotic pea crabs (Pinnotheroidea), are both old lineages that branched away from others early in the evolution of the group.

A DIET OF MUCUS

Each gall crab species is typically associated with certain species of its own favourite genus or family of stony corals, which they have specially evolved to match. However, they all seem to find coral mucus to be a rich and sustaining diet.

SPINY-CLAWED DEEP-SEA CRAB
Corycodus disjunctipes

THESE TINY BUT SPECTACULAR deep-sea crabs belong to the Archaeobrachyura, a major division of 'primitive' crabs that first evolved and dominated the sea floor during the Jurassic and Cretaceous Periods, from around 200 to 65 mya. The last two pairs of legs are positioned over the back of the shell and have small, reflexed dactyli that help to hold and carry items of camouflage. *Corycodus* species are still only known by very few specimens, so their biology and ecology is still unknown. There are seven described species; four are only known from waters off southeastern Africa, two from the Philippines, and one from the Caribbean Sea.

UNIQUE BREEDING ADAPTATION

Most marine crabs produce vast numbers of tiny eggs that are broadcast widely, although only a few will survive. *Corycodus* species, on the other hand, appear to brood fewer than ten large eggs (each egg about 20 per cent of the width of the adult body). Even more remarkable is that the brood chamber of one female was found to have a mix of five early-stage eggs, two newly hatched zoeae, and one larger zoea all at once. Brachyurans typically have one batch of eggs at a time, so it seems that *Corycodus disjunctipes* may be unique amongst known crabs in being able to continue maternal care of hatching zoea for up to three generations.

FAMILY:	Cyclodorippidae
DISTRIBUTION:	Only known from the western Indian Ocean off Mozambique, East Africa
HABITAT:	Trawled from 200–300 m (650–1,000 ft) depth; open, firm bottom
FEEDING HABITS:	Presumed to be a predator of tiny fish and other small soft-bodied invertebrates
NOTES:	The spinous claws suggest it may be an ambush predator that pierces its prey and holds it secure while it feeds
SIZE:	To about 9 mm (3/8 in) carapace width

A MATTER OF SIZE

This crab's fearsome chelae look formidable, but not so scary when it is considered that the whole crab is only 6.5 mm (1/4 in) across the shell. The enormously swollen pleon is needed to protect its clutch of very large eggs.

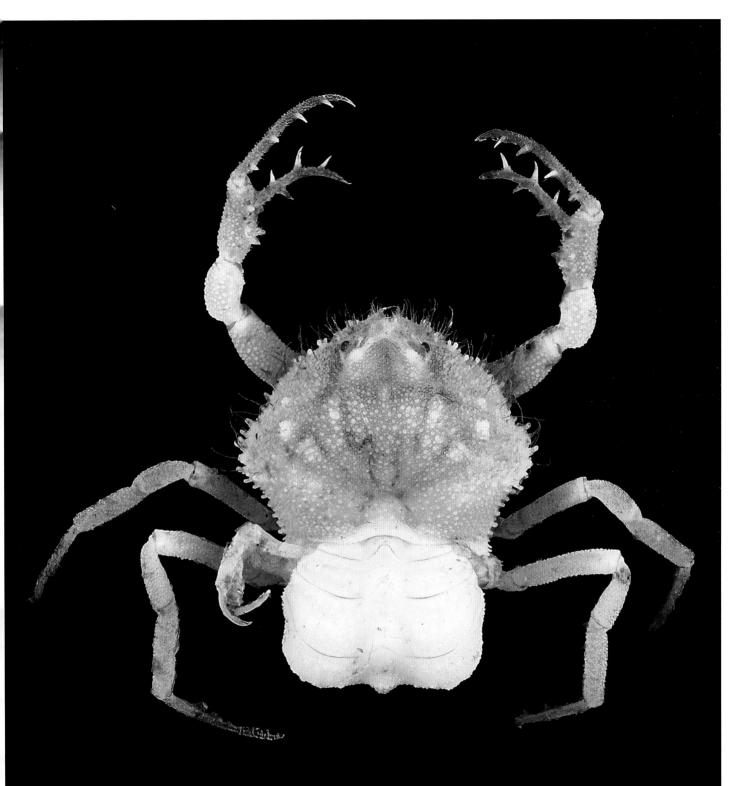

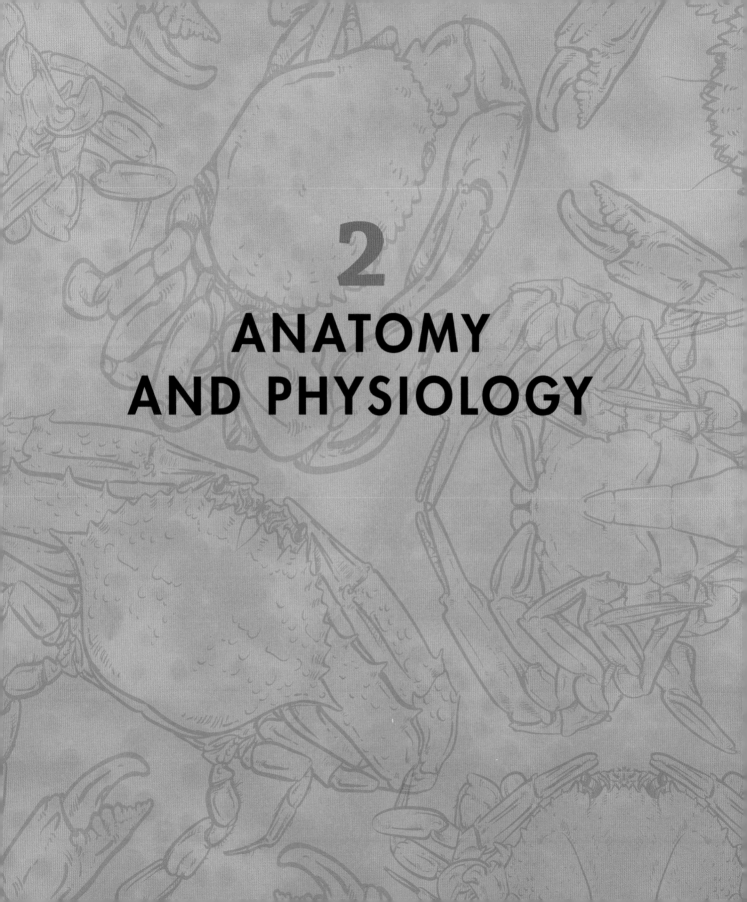

2
ANATOMY AND PHYSIOLOGY

HOW CRABS ARE PUT TOGETHER: EXTERNAL ANATOMY

BRACHYURANS, LIKE OTHER ARTHROPODS, have a hard cuticle called an *exoskeleton*, which supports the body, provides mechanical strength and protection from the environment, and affords an internal surface for muscle attachment. The exoskeleton has three outer layers: the outermost *epicuticle*, which is thin, waxy, and the main waterproofing barrier; and a thick, structural *procuticle* consisting of an outer *exocuticle* and an inner *endocuticle*. The two procuticle layers are hardened by both calcite and calcium carbonate.

The crab body is made up of a series of jointed modules, called *somites* or *segments* – a bit like earthworms, but much more sophisticated! During development, these somites are fused together in various ways to form specialized functional units (*tagmata*), namely the *cephalon* (head), the *pereion* (thorax) and the *pleon* (the abdomen or tail). The joints between the original somites can become indistinguishable in the adult, most obviously in the head. The segments of the pereion and pleon all retain a pair of variously modified jointed appendages (although some may be lost as adults). These appendages are grouped together according to their function, such as respiration, feeding, movement or reproduction. In crabs, the conjoined head and thoracic somites (*cephalothorax*) are covered by a large, thick, protective shell (*carapace*).

BELOW: There is a huge variety in carapace shapes. This spindle crab, *Ixa inermis*, takes width to the extreme, and could be mistaken for a piece of inedible broken coral.

OPPOSITE: As with human anatomy, crab limbs and carapace structures have their own highly specific terminology used for describing them.

ANATOMY AND PHYSIOLOGY

KEY PARTS OF A CRAB

FROM ABOVE

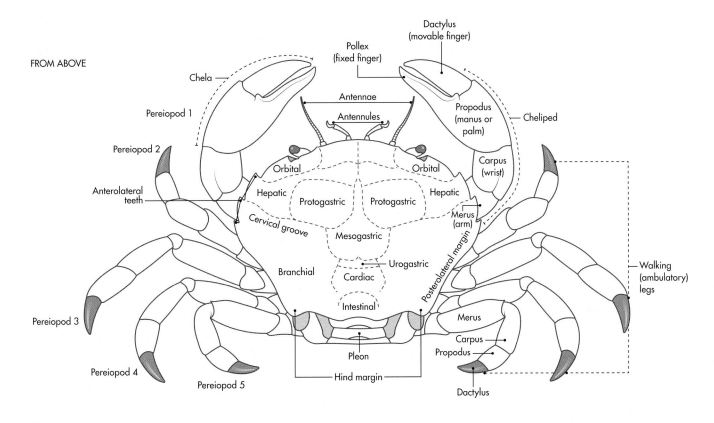

FROM BELOW

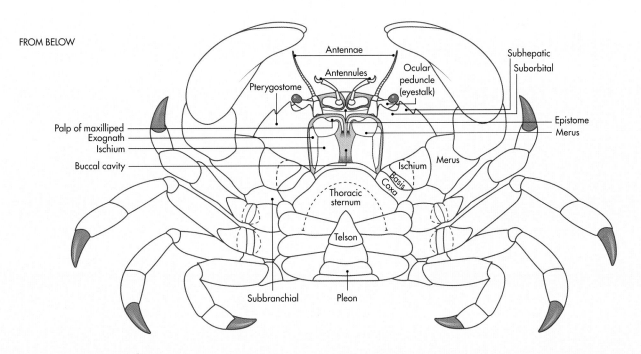

HOW CRABS ARE PUT TOGETHER: EXTERNAL ANATOMY

The first actual structure of the head is called the *acron* – this is not a true somite, but a forward extension of the brain that bears the eyes. The acron is followed by five fused body somites from which the *antennules*, *antennae*, *mandibles* (a pair of molar-like teeth) and two other feeding appendages (*maxillae*) emerge.

The antennules typically fold into a cavity under the front of the shell, between the eyes. Their main function is to detect food odours by flicking and tasting the water (see page 106). The base of each antennule also has a special organ called a *statocyst*. Each statocyst consists of fluid-filled semicircular canals lined with tiny sensitive *setae* (hairs) that detect changes in movement and keep the crab properly orientated, just like organs in the inner ear of vertebrates.

Each antenna typically has a large supporting basal segment, followed by a series of smaller, and increasingly slender, segments that develop into a sensory *flagellum*. At the base of the basal segment is an opening from the *antennal* (or *green*) *gland*. This small internal gland is located at the base of each eyestalk, covered by connective tissue and a bladder sac, and acts much like a human kidney, regulating the dilution of tissue salts. The antennae of Corystidae species, which bury into soft sediments, have developed a long, stiffened 'antennal tube', formed by interlocking thick setae on the inner edge of each antenna, through which they pump away respiratory water. The purpose of these antennal tubes is not fully understood – they may help the crabs avoid disturbing the muddy sand in which they hide; or they may direct away any chemical signals the crabs produce, so as not to alert predators to their exact position.

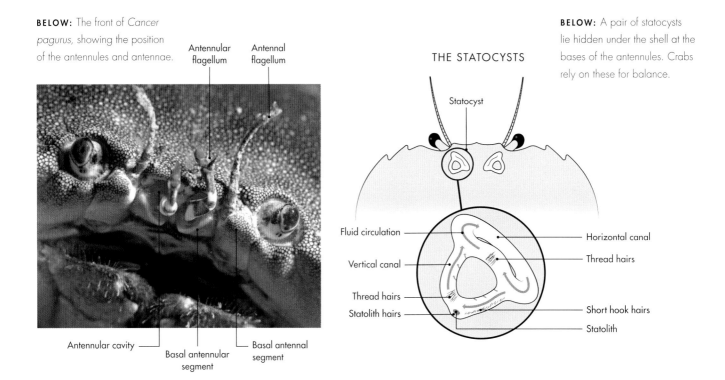

BELOW: The front of *Cancer pagurus*, showing the position of the antennules and antennae.

BELOW: A pair of statocysts lie hidden under the shell at the bases of the antennules. Crabs rely on these for balance.

64   ANATOMY AND PHYSIOLOGY

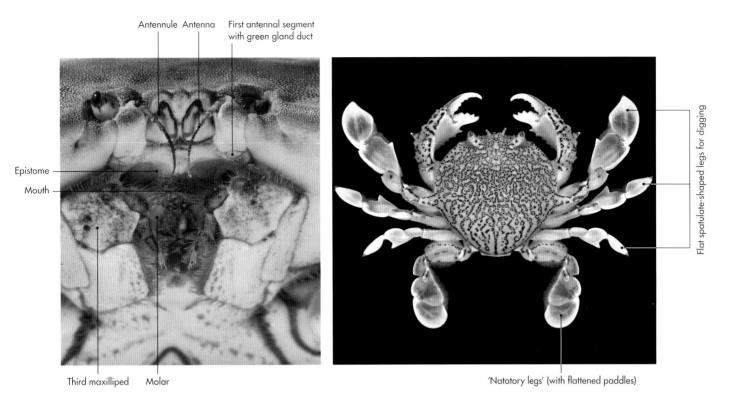

The mandibles are calcified crushing or cutting structures that sever off pieces of food before they are passed into the *gastric mill*, a specialized, stomach-like organ behind the mouth. The two pairs of *maxillae* are flattened, leaf-like appendages used for holding and manipulating food through the mandibles and into the mouth. The second maxilla also has a lateral flap called a *scaphognathite*, used to pump water through the gill chambers.

The thorax comprises eight somites, each bearing limbs used for feeding and locomotion. The first three pairs of thoracic appendages are called *maxillipeds* (from the Latin for 'jaw feet'), and are primarily involved in food manipulation. Along with other Decapoda, brachyurans have ten legs (*pereiopods*) arranged in five pairs. The last four pairs are typically used for locomotion, and referred to as ambulatory or walking legs, although they can be modified for other purposes in some families, such as swimming (Portunidae), digging (Matutidae) or carrying (Dorippidae and other families). The first pair, however, are highly modified into claws (*chelipeds*). Typically, each pereiopod consists of six segments joined by a flexible section of cuticle and a paired ball-and-socket joint. The movement of each segment is restricted to a single plane, but the direction of each plane is perpendicular to that of its neighbour (except for the basis–ischium joint, which pitches the axis of the leg forward), enabling a wide range of movement.

ABOVE LEFT: Claws pass food to the mouthparts, where it is grasped by the outermost maxillipeds (literally 'jaw feet') and fed into the molars that cut and crush it into small pieces.

ABOVE: A flower moon crab (*Matuta planipes*) has spatulate legs that allow it to dig rapidly and vanish into the sand. It is also an excellent swimmer due to its paddle-like last pair of legs.

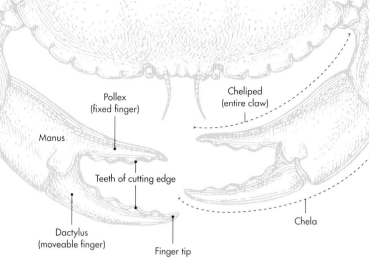

PARTS OF A CLAW

FEELING THE PINCH: CLAWS

The terms 'crab' and 'claw' are almost interchangeable in most people's minds, and for good reason – claws have been fundamental to the successful evolution of brachyurans. Claws are used for a variety of purposes: first and foremost for feeding, but also for defence, attack, display and grooming. The scientific term for a whole claw is *cheliped,* derived from Latin and meaning 'clawed foot'. The *chela* itself (pincer) is the bit that bites! The chela is formed by the propodus swelling into a prominent palm (*manus*) and growing an extended fixed finger (*pollex*); the dactylus then closes against the pollex to form the gripping device.

Chelipeds come in a great range of shapes and sizes, each enabling different combinations of speed and power to suit different purposes. Generally, heavy, muscle-filled chelipeds move more slowly than lighter, weaker ones (with exceptions). Crabs that prey on sedentary, hard-bodied prey, such as bivalve molluscs, have the most powerful chelipeds, with the highest mechanical advantage for crushing. Species exploiting more mobile prey, such as polychaete worms and fish, have lighter but longer chelipeds. The 'boxer' or 'anemone' crabs uniquely possess slender, symmetrical chelipeds that are smaller and shorter than the other pereiopods; their sole purpose is to grip live sea anemones, which the crabs use for food gathering and defence.

LEFT: A box crab (*Calappa calappa*) eating a marine snail. The peg-like tooth on its right claw is used to break open the shell and expose the snail within.

OPPOSITE: Crab claws come in a huge variety of shapes and sizes that reflect the lifestyle and diet of their owners – some crush, some scrape, and some attack.

CLAW TYPES

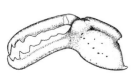
Symethis corallica (Raninidae)

Ranina ranina (Raninidae)

Dromidiopsis tridentata (Dromiidae)

Sakaila africana (Aethridae)

Tokoyo eburnea (Leucosiidae)

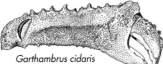
Garthambrus cidaris (Parthenopidae)

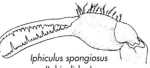
Iphiculus spongiosus (Iphiculidae)

Arcania septemspinosa (Leucosiidae)

Calappa philargius (Calappidae)

Ashtoret lunaris (Matutidae)

Corycodus disjunctipes (Cyclodorippidae)

Lybia tessellata (Xanthidae)

Pilumnus granti (Pilumnidae)

Paramedaeus megagomphios (Xanthidae)

Liomera nigrimanus (Xanthidae)

Thalamita sima (Portunidae)

Lupocyclus philippinensis (Portunidae)

Platychirograpsus spectabilis (Glyptograpsidae)

Percnon affine (Percnidae)

Sarmatium unidentatum (Sesarmidae)

Guinusia dentipes (Plagusiidae)

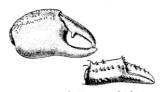
Baruna socialis (Camptandriidae)

Tasmanoplax latifrons (Macrophthalmidae)

Macrophthalmus grandidieri (Macrophthalmidae)

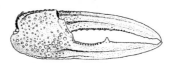
Uca dussumieri (Ocypodidae)

Scopimera inflata (Dotillidae)

HOW CRABS ARE PUT TOGETHER: EXTERNAL ANATOMY

LEFT: The male tetragonal fiddler crab (*Gelasimus tetragonon*) has one very large claw for courting females and fighting rivals. The other tiny claw is their only tool for feeding. Its long eyestalks give it 360° vision on sandy mud flats.

While the claws of many crabs look the same left and right, predatory crabs are typically *heterochelous,* meaning their chelae are different in size and shape. The larger chela usually has a 'crusher' role, and has blunt molar-like 'teeth' for breaking the shells of hard-bodied prey; the smaller chela is more designed for cutting, and usually possesses sharper, narrower teeth, and pointed tips to the 'fingers'. The two chelae thus complement each other to manipulate, crush and tear tissue from prey. A most remarkable example is found in box crabs of the genus *Calappa.* The larger right claw has a specialized supplementary tooth on the outside of the base of the lower finger that is used like a tin opener on gastropod snails (see page 66).

The chelae are also a very good guide to the feeding ecology of crabs. Crabs that consume tough plant material typically have heavy chelipeds with pointed fingertips, and dentition designed to tear and cut leaves and seedlings. Such crabs predominate in mangroves and terrestrial forests, where vegetation is the main food source. Spoon-tipped fingers are generally used for feeding on detritus, scooping up coral mucus or other soft foods, gripping filamentous algae, or scraping algae off coral rock. Scalloped fingertips are even more effective for scraping, be it encrusting algae off rocks and bark, or thin layers of leaf tissue. Such crabs usually have relatively weaker, and often more slender, chelae with less muscle mass.

The shape and size of chelipeds can also differ between the sexes, and mature male chelipeds are often larger and more robust than those of females. This is usually because the males use them for fighting other males, and for holding females 'captive' (protected from other males) before mating. Sometimes the differences can be extreme, as in the case of intertidal fiddler crabs. Female fiddlers have two small, slender claws for feeding on the mud, while the male has a dramatically enlarged, flat and colourful claw for displaying to females and to battle other males. Pure showmanship!

CRABS' LEGS AND HOW THEY MOVE

Crabs are famous for walking sideways, and although this is very common, the complicated way that the leg segments hinge together also allows many species to walk forwards, backwards or diagonally as needed. The problem of having multiple pairs of legs is the high degree of coordination required to keep them from bumping into each other. This is not so difficult when just extending and contracting sideways, but walking forwards rapidly requires much more precise control. Coordination depends on quick nerve conduction times between the muscles in the leg segments, and the levator and depressor muscles located in the cephalothorax, which provide the main propulsive and recovery strokes. A species of sandy beach ghost crab (*Ocypode*) holds the world land speed record for crabs, attaining velocities of around 4 m (13 ft) per second while running sideways! It achieves this by both leaping into the air, and periodically turning 180 degrees, thus resting the legs doing the pulling for those that were only pushing.

Crabs' legs are not only good for walking – some are modified for fast swimming, some for clinging onto coral, some for carrying camouflage on their backs, and some for rapid digging into the sediment. For example, swimming crabs of the family Portunidae have the last segment of the last pair of legs flattened into paddles. In several families, the last two pairs of legs may also end in small claws for gripping pieces of hollowed-out sponge, or for holding leaves, shells or even other invertebrates on their backs.

ABOVE: An Atlantic ghost crab (*Ocypode quadrata*) from South Carolina. A species of ghost crab holds the land speed record for a running crab.

BELOW: The last pair of legs of some members of the family Hexapodidae – such as the burrow-dwelling *Hexaplax aurantium* – have for some reason been reduced to tiny vestiges.

HOW CRABS ARE PUT TOGETHER: EXTERNAL ANATOMY

LOSING AND REPAIRING LIMBS

Autotomy (self-cutting) is akin to a lizard dropping its tail, and usually happens because of an injury, or to escape an attack. It involves the separation of the cuticle, blood vessels and nerves at the basis-ischium joint at the bottom of a leg or claw. Autotomy is a reflex response enabled by specialized levator muscles that basically pull each other apart and fracture the cuticle. It is made possible by a unique double-autotomy membrane that extends across the limb base. As the break occurs, blood pressure in the haemocoel causes the membrane to balloon outwards to immediately close the hole. Regeneration of a new limb occurs during the next few moults, with a folded limb-bud first emerging from the autotomized joint. Sometimes a claw or other limb can just be damaged without being lost. In this case repair also occurs during the next moult, with special cells capable of regeneration being recruited to where they are needed.

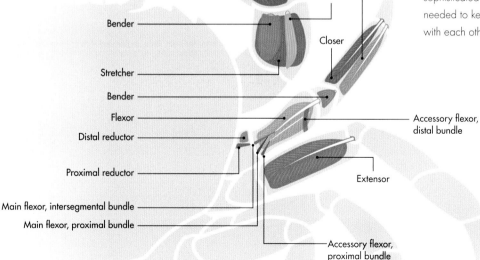

ABOVE: Sometimes a break triggers the wrong repair mechanism, especially in claws. Repair cells begin making a new, non-functional, claw wherever the break occurs, producing grotesque results, as in this brown crab (*Cancer pagurus*) found off the coast of Cornwall.

LEFT: A complex arrangement of muscles is needed to move a crab's jointed limbs, and a sophisticated nervous system is needed to keep them coordinated with each other.

THE PLEON

The pleon consists of seven segments: six true somites and a terminal *telson* (a special extension of the last somite that lacks appendages or internal organs). While this is always the normal developmental pattern, adult crabs in some groups can have somites variously fused together, so there may appear to be fewer. The pleon is a relatively simple muscular organ containing nervous and vascular blood supply. In most adult females, somites 2 to 5 each have a pair of ventral appendages called *pleopods* (often called swimmerets in other decapods, but no longer used for locomotion in crabs). These pleopods have long setae to which the fertilized eggs are attached. By waving the pleopods, and opening and closing the pleon, mother crabs ensure that their developing eggs are nurtured with fresh, oxygenated water. In adult males, however, only the first two pairs of pleopods remain, but they are highly modified for mating (called *gonopods*). Female pleons are typically broadly expanded for their egg-carrying role, and tend to cover the thoracic sternum. Males mostly have narrow abdomens that lock into a sternal cavity using a special 'press button' mechanism.

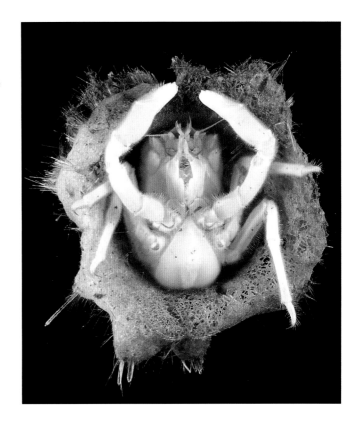

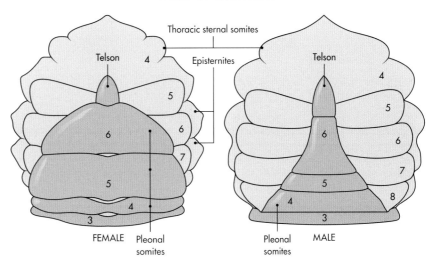

PARTS OF THE PLEON

ABOVE RIGHT: *Dicranodromia chenae* carries sponges on its back using claw-tipped legs. As with most females, the pleon covers the sternum in a dome to protect developing eggs.

RIGHT: Female pleons are broadly oval; male pleons are much narrower, and usually lock into the thoracic cavity to protect the genitalia. There are typically six somites, with the first often narrow and mostly hidden under the back of the carapace.

REPRODUCTION

One of the most significant events in crab reproductive evolution was the change from external to internal egg fertilization. In the 'primitive' podotreme and archaeobrachyuran families (see page 14), each female genital opening (*gonopore*), from which the eggs are extruded, is positioned on the coxa of the second walking leg. Male sperm is stored within special receptacles in the thoracic sternum, called *spermathecae*, which have no internal connection to the ovaries. The sperm is released from the spermathecae to fertilize the eggs externally within the enclosed space under the female pleon.

ABOVE: Australian blue swimmer crabs (*Portunus armatus*) mating. The male inserts his slender pleon beneath that of the female, allowing his paired gonopods to enter her sternal vulvae.

BELOW: In podotremes, eggs are extruded from the ovary via the oviduct and emerge from gonopores near the base of the second leg; sperm storage is in separate seminal receptacles (spermathecae), and fertilization is external. In eubrachyuran crabs, the ovary is connected to spermathecae by way of oviducts and eggs are fertilized internally.

THE FEMALE REPRODUCTIVE SYSTEM

PODOTREMATA | EUBRACHYURA

- Ovary
- Oviduct
- Coxa
- Spermatheca
- Female gonopore
- Aperture of spermatheca
- Vulva
- Vagina
- Thoracic sternum
- Ovary
- Spermatheca
- Oviduct
- Pereiopod 3

ANATOMY AND PHYSIOLOGY

THE MALE GONOPODS

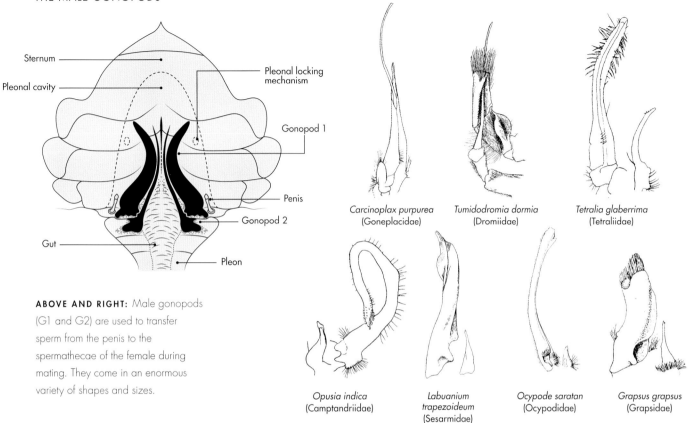

ABOVE AND RIGHT: Male gonopods (G1 and G2) are used to transfer sperm from the penis to the spermathecae of the female during mating. They come in an enormous variety of shapes and sizes.

The majority of crab families, however, belong to the Eubrachyura, in which the female gonopore is found in the middle of the sternum, on sternite 6. The sperm are inserted directly into the sternal vulval opening, and sperm storage is within seminal receptacles (spermathecae) that are connected to the ovaries via the oviducts. Fertilization occurs inside these receptacles, which have a secretory cell-lining that keeps the sperm healthy and viable until the eggs are released.

When podotremes mate, the male uses paired copulatory appendages (gonopods) to transfer sperm into the females' spermathecae; in contrast, eubrachyurans are unique in transferring sperm into the seminal receptacles through the 'vulva' and 'vagina'. In many female crabs, the vulva is calcified shut and mating is only possible when the shell is still soft from moulting. Others, however, have developed a hinged *operculum* that opens to allow mating at any time.

The first gonopod (G1) takes the form of a cylindrical tube enclosing an ejaculatory duct (or sperm channel). The G1 of podotremes is partially open along its length, and has only one large basal aperture (foramen) for the introduction of both the penis and G2. In eubrachyurans the tube is fully closed, but there are two basal openings: one for the insertion of the G2, and one for the entry of the penis. In all crabs, the G2 fits into the internal channel formed by the folds of the G1, and serves to either guide the sperm along the G1, or as a piston to pump the sperm up the channel into the vulva.

The male G1 is sometimes fairly simple, but more commonly has variously shaped (sometimes bizarre) tips that fit the female vulval aperture like a matching lock and key. Sperm transfer can occur over a protracted period, and this gonopod lock probably helps to hold the couple together. Indeed, the specific shape of the male gonopods has long been used as a reliable way of recognizing even closely related species.

HOW CRABS ARE PUT TOGETHER: EXTERNAL ANATOMY

THE EYES

A crab's eye is typically formed by a moveable eyestalk, ending in a rounded, pigmented cornea. Crabs that live in open, flat environments often have very long eyestalks, which can act like periscopes (see page 108). Special wrap-around corneas also provide a panoramic, and often 360-degree, field of view, so crabs sitting in a tide pool, or on the edges of a burrow, can easily spot potential predators. By contrast, good distance perception in highly structured three-dimensional habitats (such as rocky shores or mangrove forests) requires the binocular comparison of two separate images (*stereopsis*), just as in humans. Thus, the best design is widely separated eyes on short eyestalks (the grapsid-sesarmid design). In contrast, eyes may be substantially reduced or even lost in many cave-dwelling and deep-water species that live in perpetual darkness (for example, species of Bythograeidae that can be found around hydrothermal vents; see page 117).

ABOVE: The Atlantic mangrove root crab (*Goniopsis cruentata*), like most of the Grapsidae family, has short eyestalks at the outer edges of the carapace, giving it stereoscopic vision and distance perception.

LEFT: This deep-water hydrothermal vent species, *Gandalfus yunohana*, lives in total darkness – its eyes have become shrunken, although not completely useless.

GROWTH AND MOULTING

Like all arthropods, crabs do not increase in size continuously, but grow by increments as they moult their old shells. When still young, all their spare energy is channelled into growing quickly, and the time between moults is relatively fast; as they increase in size and become mature, this inter-moult period slows down. This is partly because it takes longer to build enough body mass to need a new shell, and partly because they are putting more energy into mating and reproduction. Growth rates are also governed by the environment – cooler water and limited food increase the time between moults. The number of moults a crab goes through depends on its lifespan, which can vary greatly. Some tiny tropical false spider crabs (Hymenosomatidae) probably live for less than a year, while the giant temperate-water Japanese spider crab (*Macrocheira kaempferi*) is reputed to live for over 100 years!

Moulting (*ecdysis*) is a complex process that involves shedding the entire exoskeleton, including the lining of the stomach and hind-gut. In preparation for moulting, calcium from the exoskeleton is reabsorbed and put into solution in the blood, where it is transported to cells, tissues and other organs for storage until it is needed to harden the new shell. Finally, the *hypodermis*, the epidermal layer of cells that will secrete the new exoskeleton, separates from the endocuticle, thus freeing the crab inside from its old shell. During ecdysis, the shell breaks open along a variety of weakly calcified suture lines (*lineae*) around the edge of the carapace. Typically, the hind margin of the carapace lifts away, and the new crab slowly backs out of its old shell. It immediately starts absorbing water, increasing its blood pressure as much as five-fold, and ensuring the new cuticle expands to its fullest extent within only one or two hours. Hiding in holes or under rocks for a day or two is a necessity until the crab can use its stored calcium to harden its exoskeleton, and become able to feed and protect itself again.

BELOW LEFT: A leucosiid pebble crab (*Arcania undecimspinosa*), just completing a successful moult. This is a dangerous time, as its new shell is soft, making it especially vulnerable to predators.

BELOW: A shag-pile crab (*Serenepilumnus kukenthali*). The setae (hairs) that cover crabs like this are extensions of the cuticle and must also be renewed with each moult.

WHAT'S HIDDEN INSIDE: INTERNAL ANATOMY AND PHYSIOLOGY

THE DIGESTIVE SYSTEM

The brachyuran gut is essentially a simple tube running through the middle of the body from the mouth to the anus (at the end of the telson). It is divided into three distinct sections: the *fore-gut*, *mid-gut* and *hind-gut*. Food is first manipulated and shredded by the chelipeds, then passed through to the mouth by the actions of the third to first maxillipeds, and the two pairs of maxillae. The pair of external mandibles cut and crush the food before it enters the mouth, but the process of grinding it into digestible-sized particles takes place in the fore-gut via the action of toothplates in the gastric mill.

When food enters the fore-gut it is initially mixed with digestive fluids before being ground. In carnivores, the gastric teeth are fewer and flatter, and are shaped to pulverize meat. In contrast, herbivorous crabs typically have sharp cusps and raised ridges shaped to cut and grind vegetable material into fine fragments – this maximizes the surface area exposed to cellulase and other digestive enzymes. Plant carbohydrates, cellulose and hemicellulose, are a major source of energy, but are difficult for most animals to digest. It used to be thought that crabs needed bacteria and fungi to 'pre-digest' leaves before they were eaten, but it has now been proven that crabs themselves produce cellulases in their fore-gut. Thus, terrestrial crabs (such as gecarcinids and sesarmids) that rely on leaves and other vegetable material are able to thrive on a diet normally considered to be very low in nutrients.

Food transit time through the gut is enormously variable, depending on the species and the type of food. For example, a herbivorous gecarcinid like the Christmas Island red crab (*Gecarcoidea natalis*) takes about 12 hours to digest its meal, and the giant mud crab (*Scylla serrata*) likewise takes about 12 hours to process a meal of prawns, fish and bivalve molluscs; however, a much smaller swimming crab of the same family (*Liocarcinus puber*) eating the same diet may take up to 24 hours, or as long as 72 hours if fed brown algae.

Recent experiments on the European green crab (*Carcinus maenas*) have added an extra dimension to our understanding of crab nutrition. These crabs are also capable of feeding through their gills! While all gills are capable of respiration, the posterior gills tend to have a thicker epithelium and play an important role in ion regulation (maintaining internal chemical stability). It has now been shown that amino acids important for nutrition pass through the gill membranes directly into the 'blood' (*haemolymph*). This has been shown to happen by passive diffusion, but crabs are likely to actively pump such nutrients across the gill surfaces. While a crab could not just live on water, nevertheless it seems that this avenue provides an important dietary supplement, and could be critical in helping crabs survive through lean times.

BELOW: An Atlantic rock crab (*Cancer irroratus*) feeding on a scallop. Crab diets can be enormously varied.

UNDER THE BONNET

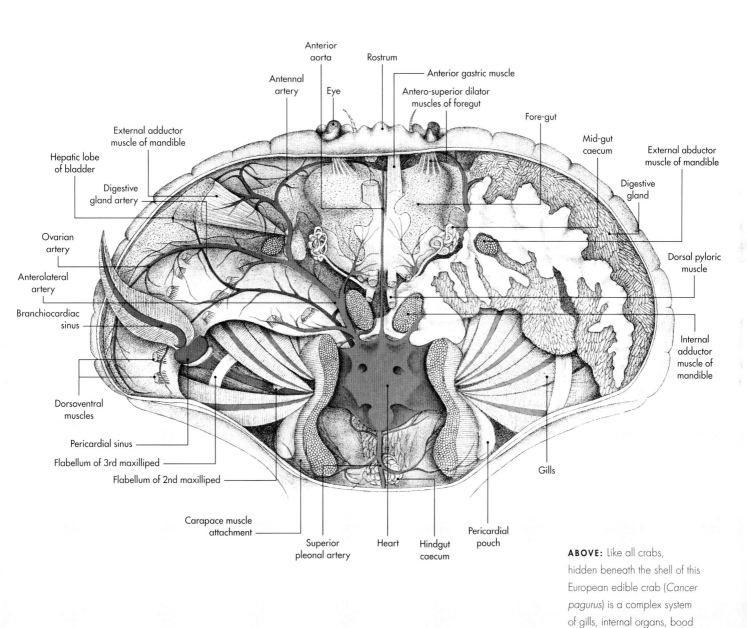

ABOVE: Like all crabs, hidden beneath the shell of this European edible crab (*Cancer pagurus*) is a complex system of gills, internal organs, bood vessels and nerves.

WHAT'S HIDDEN INSIDE: INTERNAL ANATOMY AND PHYSIOLOGY

THE CIRCULATORY SYSTEM

The evolution of a sophisticated arterial system was crucial in enabling crustaceans to become large and highly mobile. Brachyurans have an 'incompletely closed' circulatory system that is much more complex than has been assumed in the past. The haemolymph (blood) is distributed by a heart and arteries, but unlike in vertebrates, a sinus system collects and returns oxygenated blood for re-circulation rather than a closed system of veins. Crab's blood is copper-based hemocyanin, giving it a translucent blue colour, rather than the red of vertebrate haemoglobin.

The heart lies within a blood-filled pericardial cavity where it is suspended by elastic ligaments attached to the pericardial wall. The oxygenated haemolymph in the pericardium enters the heart through a number of valved *ostia* (slit-like holes) in the heart wall. Heart contraction pumps the blood through a series of five arterial systems (each with valves to prevent back-flow) to all parts of the body. The arteries branch and become increasingly small as they penetrate the tissues, finally ending in vessels 2 to 10 microns in diameter (comparable to vertebrate capillaries). After each contraction, the pericardial ligaments re-expand the heart.

Heart rate is controlled by nerve signals from the cardiac ganglion and the central nervous system, or by neurohormones acting on the cardiac muscle, so high blood pressures can be produced quickly if needed. Variation with activity can be marked (e.g. from 40 to 200 beats per minute in *Carcinus maenas*), but stress factors, including hunger, thermal stress, anoxia (slows with decreasing oxygen), carbon dioxide levels, preferred substrate type, and exposure to heavy metals and other pollutants, will also affect heart rate.

BELOW: Stress factors affect heart rate. *Neosarmatium integrum* (left) lives intertidally in tropical mangroves, where it endures high temperatures and low oxygen. The grapsid rock crab (right) would be severely stressed by oil pollution washed ashore.

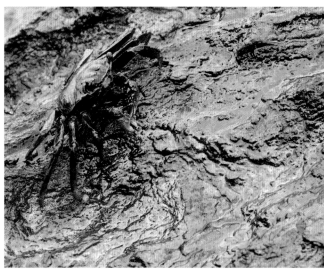

THE CIRCULATORY SYSTEM

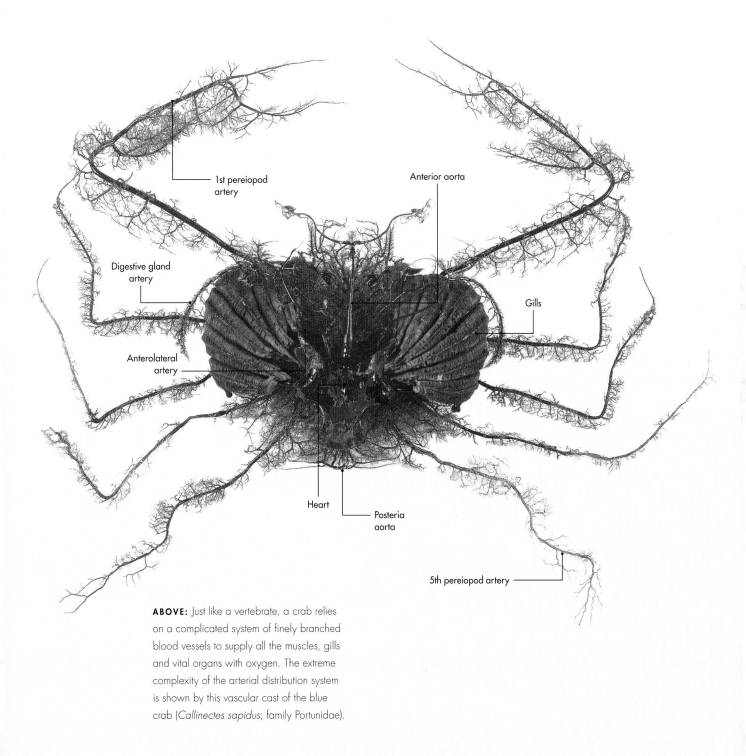

ABOVE: Just like a vertebrate, a crab relies on a complicated system of finely branched blood vessels to supply all the muscles, gills and vital organs with oxygen. The extreme complexity of the arterial distribution system is shown by this vascular cast of the blue crab (*Callinectes sapidus*; family Portunidae).

GILLS AND RESPIRATION

Gills are the main organ of respiration in most crabs and have a rich blood supply. They are formed as special outgrowths from the bases of the legs and feeding appendages, and are contained in a special branchial chamber on each side of cephalothorax. Crabs have *phyllobranchiate* gills, which means that they are made up of many thin, leaf-like filaments (*lamellae*) that project from a central shaft. This arrangement creates a large surface area to absorb oxygen from the water as it flows between the lamellae. The gills are multifunctional and they underlie many physiological functions, such as osmoregulation (tissue salt balance), the maintenance of calcium in the hemolymph, and active excretion of nitrogen as ammonium.

Gills need lots of water to function properly, so many terrestrial crabs have also developed a highly vascular lining to the branchial chamber. This acts much like a lung by absorbing oxygen directly from the atmosphere. Such *bimodal respiration* is highly developed in the Australian amphibious air-breathing crab *Austrothelphusa transversa* (Gecarcinucidae), which is able to switch the blood supply from the gills to the branchial 'lung' as soon as the crab emerges from the water.

ABOVE LEFT: *Austrothelphusa transversa* has a remarkable ability to survive in desert conditions by breathing air through a 'pseudo lung'. In droughts, it plugs its burrow with earth and goes into a dormant state, living off fat stores.

ABOVE: Sand bubbler crabs such as *Scopimera inflata* (Dotillidae) have enlarged oval patches of thin, decalcified cuticle on the legs. These 'gas windows' allow respiration directly through the shell while the tide is out.

Gills require oxygen-rich water to function, which may not be readily available in all environments. In mangrove swamps, for example, the high biological oxygen demand of the rich organic muddy sediments means that burrow water can quickly lose oxygen and gain carbon dioxide. Two groups of mangrove crabs (Sesarmidae and Varunidae) have evolved a remarkable system to cope with these conditions – they recycle and re-oxygenate the water. A complex series of grooves channel water away from the branchial aperture at the top of the mouth, draining it downwards in a thin film across myriad rows of short curved setae that cover the *pterygostome* (side walls of the carapace). This film of water absorbs oxygen from the air, and then is collected and channelled back to the gills via openings at the bases of the chelipeds.

THE NERVOUS SYSTEM AND COGNITION

Crabs have complex lives, complex anatomy and complex behaviour, so they need a complex brain and nervous system that allows them to perceive, and respond to, their environment. Nervous control is evident in every aspect of their lives, including movement; highly sophisticated vision with complex depth perception; vibrational 'sound' inputs through special sensory setae, and from special organs at the bases of their legs; smell (olfaction) through highly sensitive chemo-sensory antennules; control of moulting and nervous control of hormone secretions; and even colour changes.

Compared with other decapods, the brachyuran central nervous system is strikingly compressed because of the restrictions imposed by carcinization. It consists of a large supraesophageal ganglionic mass (the *cerebral ganglion* or 'brain'), located in the mid-line under the frontal region. Major nerves radiate from this to the eyes, antennae and antennules. Posterior to the main brain lies the *thoracic ganglion*, from which nerves radiate to all the other thoracic appendages.

Studying the development of different parts of a crab's brain and central nervous system can reveal evolutionary adaptations to a variety of environmental conditions. For example, well-developed optic *neuropils* (dense networks of nerve fibres) within the eyestalks indicate a good ability to process visual input; similarly, distinct olfactory neuropils suggest enhanced chemo-reception. The brain of the European green crab (*Carcinus maenas*) shows both enhanced visual and olfactory development – especially useful in the turbid or dark marine environments it inhabits – but the reduced development of the tritocerebrum suggests that mechano-sensory information from the antennae is somewhat less important to them, and indeed their antennae are relatively reduced compared to many other crabs.

THE CENTRAL NERVOUS SYSTEM

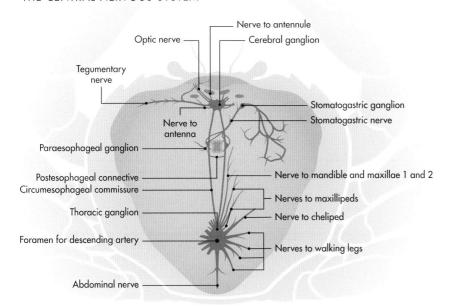

LEFT: Brachyurans have an unexpectedly sophisticated nervous system that controls almost every aspect of their lives. New studies have revealed remarkable abilities to problem-solve mazes and memorize landmarks.

CRAB COLOUR

Brachyurans have an enormous range of species-specific colours and patterns, employing them for display, mate recognition, communication and camouflage. For example, in some fiddler crab (*Uca*) species, females identify males based only on claw colour, and males recognize female neighbours by their distinctive carapace colour patterns. The colour of a brachyuran is a function of the number, type and distribution of *chromatophores* (located in the epidermis), as well as the types and quantities of pigments in the exoskeleton. Chromatophores normally each contain a single pigment, so are classified by colour: melanophores (black/brown), leucophores (white), erythrophores (red) and xanthophores (yellow). Different colours are produced by the use of a uniquely crustacean macro-molecular protein-complex, crustacyanin, which is able to modify the red carotenoid, astaxanthin, to produce any colour in the visible spectrum. Colour changes can be slow and predictably associated with becoming adult or to show reproductive readiness, or in rhythm with the seasons. They can also occur within a matter of minutes or hours, showing rapid response to fluctuations in illumination and temperature, handling or predator stress, or sexual encounters (courting or inter-male aggression).

In some tropical mangrove sesarmid crabs, the brightness and intensity of 'facial' colour bands on the males seem to be important in mate and/or species recognition. Such extravagant and colourful sexual signals are the result of carotenoid-based pigments that are only available to the crabs through their food, so the brilliance of their faces shows them to be healthy and successful, and thus a good choice for a mate! Although not well documented, it has been found in recent years that some reef crab species become brightly luminescent at night under UV light, in the same way that many corals do (see page 92). This could be another example of crabs incorporating special pigments from their food. It will make for fascinating future studies.

BELOW LEFT: A few deep-sea species, such as *Ovalipes molleri*, are spectacularly iridescent. Multi-layer reflectors in the cuticle reflect blue light parallel to the sea floor, detectable by other crabs of the same species.

BELOW RIGHT: Some reef crabs, such as *Calappa gallus*, become luminescent under UV light at night.

OPPOSITE: Each male fiddler species has its own special colour and style. From top, left to right:
1. *Tubuca paradussumieri*
2. *Austruca lactea*
3. *Tubuca flammula*
4. *Gelasimus vocans*
5. *Austruca bengali*
6. *Paraleptuca crassipes*
7. *Cranuca inversa*
8. *Afruca tangeri*

YELLOWLINE ARROW CRAB
Stenorhynchus seticornis

FAMILY:	Inachoididae
OTHER NAMES:	Arrow crab, spider crab
DISTRIBUTION:	West Atlantic Ocean: North Carolina south to Brazil; especially abundant throughout the Caribbean
HABITAT:	Coral reefs, in small caves or crevices; depths of 3–40 m (10–130 ft)
FEEDING HABITS:	Scavenger and predator
NOTES:	Used to control bristle worm populations in marine aquariums. The 5 species of arrow crab all occur in the Atlantic or eastern Pacific Oceans.
SIZE:	Up to 5 cm (2 in) carapace length; leg span to 20 cm (8 in) or more

THESE STRANGE BUT BEAUTIFUL CRABS, named for the arrowhead shape of their bodies, can be very common on the reefs where they occur. Their success is the result of the evolution of some fascinating behavioural and feeding strategies. Being nocturnal, they like to hide during the day in crevices or under ledges, and seem to particularly like associating with the branching sea anemone (*Lebrunia danae*), which no doubt affords them protection from fishy predators.

VERSATILE FEEDERS

These crabs' long, spindly legs allow them to move gracefully, if somewhat comically, over the reef, but they do have another more practical purpose. Climbing onto reef outcrops for the night, their long legs hold the crabs tall and motionless in the current, collecting passing debris on the hairy setae that cover their body. This is then cleaned off and eaten in the safety of their daily hiding places. Sometimes they also actively scavenge food and store it on their long frontal horns to be eaten later. But they can also be aggressive predators if needed, and feather-duster worms, snails, shrimps and even small unwary fish are all fair game.

NO NEED TO DRESS UP
Arrow crabs belong to a large group collectively known as spider crabs because of their long spider-like legs. Many such species decorate themselves for camouflage, but the yellowline arrow crab uses its natural striped coloration to disrupt its outline.

SLENDER-CLAWED BOXER CRAB
Lybia leptochelis

BOXER CRABS ARE SO NAMED because of their amazing habit of carrying a tiny sea anemone in each claw – tiny white pom-pom gloves with which they appear to 'box' at each other during territorial contests. These anemones are also waved and poked at potential predators. Reportedly highly venomous, small fish have been paralyzed after contact with the tentacles.

BONSAI SLAVES

It seems that the tiny anemones are kept as slaves for the crabs, which use them to mop up particles of food and organic matter that they then steal (called 'kleptoparasitism'), using their specially modified first legs. The crabs regulate the size of the anemones by effectively starving them to stunt their growth and keep them manageable. Allowed their freedom, these 'bonsai' sea anemones grow quickly to become more than 250 per cent larger. Boxer crabs hold their anemone companions with great tenacity by piercing through their base with strong spines on each claw. If an anemone is lost, the crab will cut a piece from the remaining anemone and grab it in its empty claw. The anemone then grows back whole – an extraordinary case of deliberate cloning!

FAMILY:	Xanthidae
OTHER NAMES:	Slender-clawed pom-pom crab
DISTRIBUTION:	Tropical Indian Ocean, from East Africa to Red Sea and east to Indonesia
HABITAT:	Coral reefs, typically under dead coral slabs or rocks
FEEDING HABITS:	Organic matter (detritus) from the bottom, around where it lives
NOTES:	11 species in 3 genera have been described. Several types of anemones are used: *Lybia leptochelis* prefers a species of *Alicia*, but *Triactis producta* is most common with other boxer crabs.
SIZE:	Up to 25 mm (1 in) leg span

ANENOMES AT CLAW'S LENGTH

Boxer crabs use their anemones to box at each other during territorial contests – but mostly from a distance, with no actual punches being exchanged. Their anemones are generally used for food gathering and as protection from predators.

GAUDY CLOWN CRAB
Platypodiella spectabilis

FAMILY:	Xanthidae
DISTRIBUTION:	Common in the Caribbean, but occurs from Florida south to northern Brazil
HABITAT:	Shallow-water reefs; living in zoantharian colonies; occasionally on sponges and sometimes in rubble areas
FEEDING HABITS:	Omnivore, predator
NOTES:	Some speculate that its spectacular colouring warns predators of its toxicity, but because it is so variable, it seems more likely to be a form of disruptive camouflage
SIZE:	To about 30 mm (1 1/8 in) carapace width

AS THE SCIENTIFIC NAME IMPLIES, the gaudy clown crab is certainly one of the most spectacular of marine crabs. Most crabs have relatively consistent colours between individuals, and this is usually a good indicator of species differences – each different clown crab, however, displays colours and patterns that are as specific as fingerprints. The reason remains a mystery. The colour of a brachyuran is determined by the number, type and distribution of chromatophores in the shell's epidermis, and the types and quantities of pigments in the exoskeleton. Nevertheless, a protein-complex called crustacyanin (unique to crustaceans) is able to modify the normally red pigment astaxanthin to produce any colour in the visible spectrum.

A TOXIC PAIRING

Gaudy clown crabs most commonly live commensally with zoantharians (mat anemones), and in particular, species of *Palythoa*, although they are also found living free in rubble and sponge habitats, so perhaps it is not an obligatory relationship. Adults crabs live in burrows in the base of the zoanthurian, venturing out through small holes to feed on the polyps; they are immune to the powerful palytoxin (one of the most poisonous non-protein substances known) that protects *Palythoa* species from most other predators. The crabs themselves incorporate this toxin into their own flesh, making them deadly to eat in their own right (see Chapter 5).

ONE OF A KIND

Gaudy clown crabs are members of the large family Xanthidae, the majority of which are reef specialists that play a vital role in all aspects of reef ecology. Unlike just about all other crabs, no two gaudy clowns are the same – each has its own distinctive colour and pattern.

LOPSIDED CRAB
Podocatactes hamifer

CRABS IN THE FAMILY TRICHOPELTARIIDAE are most typically trawled from deep water off the continental slope, or from oceanic seamounts 800 m (2,600 ft) or more below the surface, although *Podocatactes hamifer* is somewhat unusual by its presence in the shallower waters of the continental shelf. These crabs belong to the Eubrachyura, with fossil relatives dating from about 65 mya. The sharp spines around the carapace would have provided some protection from predators, and the long antennae would have helped locate prey and predators by sensing chemical cues and water movement.

ONE SIDE BIGGER THAN THE OTHER

Podocatactes hamifer is highly unusual amongst brachyurans for being lopsided – in mature crabs, the frontal somite of the right side is enormously enlarged compared with that of the left. The explanation seems to be that their exceptionally large claw requires an equally large block of muscle to support and move it, so the sternal compartment must expand to accommodate it. Most crabs have broad sternums that are widest across the middle, but trichopeltariid crabs have a very narrow sternum, and in *P. hamifer* it is widest at the base of the enlarged cheliped. Presumably, crabs with a broad sternum have sufficient space for the enlarged muscles needed, even if they have a large asymmetrical claw (such as intertidal fiddler crabs).

FAMILY:	Trichopeltariidae
DISTRIBUTION:	Northwestern Pacific: Japan south to Taiwan
HABITAT:	Rocky reefal environments, 50–550 m (160–1,800 ft) depth
FEEDING HABITS:	Not well known; its powerful claw suggests it is a predator of other bottom-dwelling invertebrates such as molluscs and echinoderms
NOTES:	One of only 3 species in the genus *Podocatactes*
SIZE:	Up to 25 mm (1 in) carapace length

DEEP-SEA MYSTERIES

Every expedition to explore the deep waters off our continental margins returns some fascinating creature never seen before. Billions of dollars are spent looking for life on Mars, yet a fraction of this would discover untold new treasures at the bottom of our own oceans.

PRETTY CRESTED REEF CRAB
Lophozozymus pulchellus

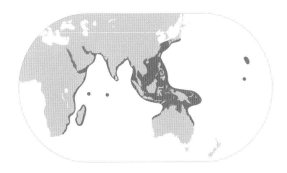

FAMILY:	Xanthidae
OTHER NAMES:	Pretty xanthid crab
DISTRIBUTION:	Widespread in the Indo-West Pacific
HABITAT:	Reefs; amongst coral and rocky rubble; to 120 m (400 ft) depth
FEEDING HABITS:	Omnivore, predator and scavenger
NOTES:	Latin *pulchellus* means 'pretty'
SIZE:	To about 20 mm (3/4 in) carapace width

CRABS OF THE FAMILY XANTHIDAE have an remarkable range of form, colour and pattern, and while they are found in a variety of habitats, they reach the zenith of their diversity on tropical coral reefs. They are commonly called 'black-fingered crabs' because the fingers of the claws are almost always at least partially black or dark brown. They occupy most niches on reefs, from grazing algae to being voracious predators of other invertebrates, and some have specialized claws for breaking mollusc shells.

BEWARE OF POISON

Colours in brachyurans are used for display, mate recognition, communication and camouflage. Xanthids also likely use their bright patterns to warn predatory fish that they are potentially deadly to eat. Numerous small reef crabs can be poisonous, but the number of xanthids tops the list. Crabs do not make their own unique toxins, but instead sequester into their flesh a variety of toxins from the environment. Saxitoxin (or 'paralytic shellfish poison') and tetrodotoxin are the two most common – powerful neurotoxins that block the sodium channels that carry messages between brain and muscles, causing rapid paralysis that extends to the diaphragm and other muscles that control breathing (see also page 192).

AGLOW IN THE DARK...
Some reef crabs become brightly luminescent at night under UV light. Fluorescence is caused by pigments that absorb one colour of light and emit it again at a longer wavelength, thus changing its colour, but the purpose is not fully understood. Crabs may incorporate special pigments from their foods, so their ability to luminesce may vary according to diet.

ROUGH-SHELLED PORTER CRAB
Dorippe quadridens

THE MOST OFTEN ENCOUNTERED shallow-water 'carrier' or 'porter' crabs belong to the family Dorippidae. Carrying behaviour was common amongst the 'primitive' crabs of the Podotremata and Archaeobrachyura, which were the first crabs to become abundant around 185 mya. The last two pairs of legs fold back over the body and are unusual in having developed pincers that the crabs use to hold various objects, both living and non-living. Hiding underneath this cover then allows them to venture out over the open substrates where they live.

PICKING A PROP

Most dorippids swap their cover as necessary, both as they grow, and depending on what is available. Species of some genera however, have more specific needs, and will only use a shell from a dead bivalve, while others will fixate on a mangrove leaf. *Dorippe quadridens* will use sponges, shell, sea urchins, starfishes or stalked barnacles; there are even accounts of them carrying around the bottom-dwelling 'upside-down jellyfish' (*Cassiopea* species), with their rear claws hooking onto the umbrella. (For more on porter crabs, see page 161.)

FAMILY:	Dorippidae
OTHER NAMES:	Sea-urchin-carrying crab, pink carrier crab
DISTRIBUTION:	Indo-West Pacific, from southeastern Africa to the Suez Canal, the Red Sea, and to Hong Kong, the Philippines, Indonesia and Australia
HABITAT:	Flat bottoms of soft to firm mud, to sand mixed with gravel, shell, rocks, coral low sponge and coral cover and sometimes seagrass. Depths typically from 1–30 m (100 ft).
FEEDING HABITS:	Omnivorous scavenger
NOTES:	Typically lies buried in the substrate during daylight hours, venturing out towards evening
SIZE:	Males to about 38 mm (1½ in) carapace width

NO RESPECT FOR SPINES

It seems unlikely that sea urchins enjoy being held prisoner, but when porter crabs lock the claws of their back legs onto an urchin, the latter has no choice but to go for the ride.

GUINOT'S AGILE REEF CRAB
Percnon guinotae

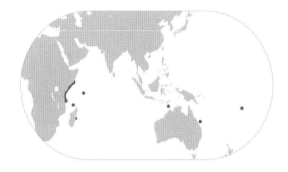

FAMILY:	Percnidae
OTHER NAMES:	Flat coral crab
DISTRIBUTION:	Widespread Indo-West Pacific; East Africa to French Polynesia
HABITAT:	Under rocks or coral blocks; shallow subtidal to about 10 m (30 ft)
FEEDING HABITS:	Herbivorous; eats macro-algae
NOTES:	Very shy; scurries away if uncovered
SIZE:	To about 25 mm (1 in) carapace width

CRAB CLAWS ARE HIGHLY VARIABLE in shape, size and purpose, and are an excellent guide to the feeding ecology of each species. Heavy, muscle-filled chelipeds often move more slowly than lighter, weaker ones, but have the power to crush shells or strip coconuts. Species of *Percnon* live in the shallow subtidal zone of rocky shores or coral reefs. These often highly colourful crabs all tend to have rounded, swollen chelae with fingers that end in strong, cusp-like tips – purpose-built to strip and scrape algae from the hard surfaces of rocks and coral.

A COLOURFUL FAMILY

This is just one of seven similar-looking species in *Percnon*, the only genus of its own family, Percnidae, whose closest relatives are amongst the intertidal shore crabs such as Sally Lightfoots (family Grapsidae). Each species has distinctive colouring, and also differs in the size and position of the spines on their limbs and carapace. Most species are found in the Indian and West Pacific Oceans, however the nimble spray crab (*Percnon gibbesi*) is established on both sides of the Atlantic (including the Mediterranean), and along the tropical and subtropical American Pacific Coast, from California to Chile.

A BODY FIT FOR PURPOSE

The flat bodies of *Percnon* species are ideal for slipping quickly into crevices, and hiding beneath rocks or coral slabs. Rows of strong, sharp spines along the frontal edge of the walking legs and claws also help lock them into their hiding places, and are a good defence against predators.

MALAYSIAN FACE-STRIPE MANGROVE CRAB
Parasesarma peninsulare

THE GENUS *PARASESARMA* represents an extremely successful group of mangrove specialists, and currently contains over 65 species, many of which, like *Parasesarma peninsulare*, have quite restricted distributions. They are common in Indo-West Pacific mangrove swamps, and they play a vital ecological role by consuming mangrove leaves and recycling plant energy and nutrients into the mud for other animals to consume.

ATTRACTIVE MATES

The 'facial' colour bands on *Parasesarma peninsulare* and related species seem to play a key role in species recognition and choice of mate. Such colours are produced by pigments that are only available through the crabs' food (see page 82), so the more brilliant their face, the healthier they are, making them a desirable mate. Facial bands only develop in fully grown, sexually mature crabs, with males tending towards intense blues and females going for brighter green shades. Members of this same group of crabs are special in another way, too. After two male crabs tussle over a female, the winner will often perform a victory dance to intimidate the loser, including defiantly stridulating one claw against the other in a crabby fist-pump (see page 158)!

FAMILY:	Sesarmidae
OTHER NAMES:	Face-band crab
DISTRIBUTION:	East and west coasts of the Malay Peninsula, south to Singapore and Batam Island (Indonesia)
HABITAT:	Relatively common species in muddy mangrove areas; active burrowers, especially within mangrove root systems
FEEDING HABITS:	Omnivorous; graze on sediment, consume mangrove leaves and fruit, and occasionally small invertebrates
NOTES:	The brightness of colours seems related to a diet rich in *Avicennia alba* mangrove leaves and fruit
SIZE:	To about 22 mm (⅞ in) carapace width

REMARKABLY ADAPTED
Face-stripe crabs belong to the diverse, primarily mangrove specialist family Sesarmidae. Sesarmids have the amazing ability to reoxygenate depleted muddy water, giving them more time to feed.

HORN-EYED GHOST CRAB
Ocypode ceratophthalmus

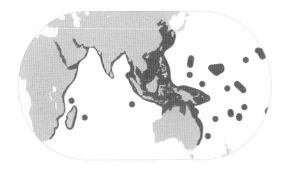

FAMILY:	Ocypodidae
OTHER NAMES:	Horn-eyed sand crab
DISTRIBUTION:	Indo-West Pacific Oceans
HABITAT:	Open sandy surf beaches
FEEDING HABITS:	Scavenger and predator; famous for eating sea turtle eggs and hatchlings
NOTES:	New research has revealed that the green to gold colour form in the Indian Ocean (shown here) is likely a new species
SIZE:	To about 40 mm (1⁵⁄₈ in) carapace width

GHOST CRABS ARE INSTANTLY recognizable for anyone who's ever visited a sandy ocean beach. Mostly tropical or subtropical in distribution, around 25 species are spread across the world. *Ocypode* is derived from the Greek *ōkys* ('swift') and *pous* ('foot'), which makes sense when they are seen in action; 'ghost crab' comes from the fact that many are pale coloured and most active at night, living in deep burrows that reach down to the water table during the heat of the day. *Ocypode ceratophthalmus* is perhaps the most common and widely distributed of all the *Ocypode* species, and is distinguished by the long, slender 'horns' protruding from the tops of its corneas – a phenomenon called *exophthalmy* (the Latin species name literally translates as 'horn-eyed'). Seven species have similar eye extensions, but none are as long and splendid.

STOMACH GROWLS

Ghost crabs are also famous for producing rasping noises with their claws so they can find each other over wide expanses of beach. Recently it has been discovered that at least some species have also evolved a loud internal 'growl', caused when the ridged plates of the gastric mill in the stomach rub together. This is used during aggressive encounters between males, and is believed to free up the claws for combat. It can be modulated in intensity, depending on the aggressiveness of the interaction.

ALWAYS MORE HOUSEWORK

Claws are not just for feeding and fighting, they are also very useful for scooping out sand from burrows that have collapsed, or been plugged against the incoming tide.

3
CRAB ECOLOGY

A MOST SUCCESSFUL GROUP...

CRABS HAVE A REMARKABLE ABILITY to exploit many environments that are off limits to most other crustaceans. Their unique body plan has allowed them to evolve remarkable physiological, morphological and behavioural adaptations to thrive in diverse, and even extreme, environments. Everywhere they live, crabs perform important ecological roles – preventing algae from overgrowing coral reefs, cleaning beaches of detritus, and aerating the mud in mangrove forests are a few of many examples. Crabs and crab larvae are also a crucial part of the diet of many sea and shoreline animals.

Crabs themselves feed in many different ways. They can be ferocious predators, scavengers or herbivores; they can also filter tiny morsels out of the water, live on coral mucus, or endlessly sieve sand and mud for meagre nourishment. However, apart from specialist feeders, most brachyurans do not have strict dietary preferences, but are opportunistic omnivores, consuming a wide range of foods, and readily cannibalize, or prey on other crab species, when the opportunity arises. A species such as the common green crab (*Carcinus maenas*) is known to prey on immobile invertebrates (mussels, cockles, oysters), attack mobile crustaceans and fish, scavenge carrion, and even consume macro-algae. Such dietary adaptability has, without doubt, been key to their evolutionary success.

CRABS AS PREDATORS

Predatory marine brachyurans are typically most diverse in intertidal and shallow subtidal habitats, because high nutrient levels coming off the land lead to an abundance and diversity of herbivorous and omnivorous prey species.

Predators have a variety of strategies. Some, like portunid swimming crabs, are fast and aggressive, catching active prey such as fish and prawns; others, like calappid box crabs, are slow and cumbersome, but armed with powerful claws to crush bivalves and gastropods. Some actively hunt, while others lie quietly in wait. For example, the specialist Indo-West Pacific mangrove forceps crab (*Epixanthus dentatus*) is an aggressive ambush predator. It will feed on almost any invertebrate that comes within range, but prefers other crabs. It is most active on nocturnal low tides, sitting still and blending in with its muddy background, waiting with claws outstretched for prey to come within striking distance. Forceps crabs have a home den amongst mangrove roots,

LEFT: The mangrove forceps crab is an aggressive ambush predator. It feeds on almost any slow-moving invertebrate that comes within range, but it particularly likes other crabs.

OPPOSITE: The montane rainforest manicou crab of Trinidad and Tobago will lie in wait to attack and kill insects, frogs, lizards, other manicou crabs and even small snakes.

but can operate within a 3-m (10-ft) radius. When feeding, they use their larger, heavier claw to crush their prey, and the other claw's long, forceps-like fingers to pick away the meat.

Another type of ambush hunter is the large amphibious manicou or Trinidad mountain crab (*Rodriguezus garmani*; Pseudothelphusidae), which lives in the high mountain rainforests of Trinidad, Tobago, Margarita and eastern Venezuela. While it preys on a range of forest animals, such as insects, frogs, lizards and other manicou crabs, it is also an important predator of several native snakes, including the clouded slug-eating snake (*Sibon nebulata*), the three-lined snake (*Atractus trilineatus*) and the vine snake (*Oxybelis aeneus*). Growing to 10 cm (3 in) carapace width, and weighing up to 250 g (9 oz), these crabs are commonly harvested for human consumption, while smaller crabs are hunted by the crab-eating hawk and other birds of prey.

Ghost crabs (*Ocypode* species) are conspicuous on ocean beaches, exposed sand banks and the sandy shores of islands and atolls in all tropical and semitropical seas. Their relatively large size, and the constant digging and maintenance of their burrows, means that they are important bioturbators (burrowing and reworking the sand), and play a key ecological role in beach food-webs. While they will predate or scavenge on whatever is available, they also target the eggs and hatchlings of sea turtles. A recent study of one of the world's largest loggerhead turtle (*Caretta caretta*) rookeries in the Cape Verde Islands, found that *Ocypode cursor* stole an average of 33 eggs per nest! Once the baby turtles emerge and make their way to the water, they are still not safe from these fleet-footed crabs. With sea turtles facing a major crisis for their survival, curbing and managing crab predation may be an important consideration.

CHEMICAL WARFARE

Many ecological communities are structured around predator–prey interactions, which often involve chemical cues for both defence and offence. A crab's antennules are invaluable in this regard – extremely sensitive organs that flick continually, helping crabs to sense chemicals in the water, much like a twitching dog's nose scans for odours in the air. Research has shown that the common blue crab (*Callinectes sapidus*) can detect as little as 1 gram of clam extract in a trillion litres of seawater. This highly developed ability may partly be due to living in murky estuarine waters, where searching by sight is often impossible. While they do like clams, blue crabs also feast on small Atlantic mud crabs (*Panopeus herbstii*), which live under rocks and around oyster beds. If blue crabs are in the area, the smaller crabs spend a lot of time hiding, but just how they know their nemesis is present has not been understood until recently. Research has shown that *Panopeus herbstii* has an equally sensitive ability to detect the distinctive smell of blue crab urine. More than 600 chemicals make up blue crab urine, but two of them – trigonelline and homarine, both common products of animal metabolism – ring warning bells for the mud crabs. Not only can they detect these chemicals, but they become even more agitated if the blue crab's last meal was mud crab! It is incredible to think that a single blue crab's consumption of one tiny mud crab, and subsequent urination, can change the interactions in the food chain of an entire estuary. This amazing sensory ability must have an enormous influence on the lives of countless different crabs, in ways we are only just beginning to comprehend.

LEFT: The Atlantic mud crab has an 'early warning system' – it can sense the presence of the predatory blue crab by 'tasting' the minute traces of urine that blue crabs secrete into estuarine waters.

OPPOSITE TOP LEFT: Ghost crabs are infamous for attacking and eating newly hatched sea turtles as they cross the beach.

OPPSITE TOP RIGHT: Many intertidal and semi-terrestrial crabs, like this Christmas Island red crab, are primarily vegetarian, and play a vital role in nutrient cycling. On Christmas Island, it is the ecological keystone species upon which the forest communities depend.

OPPOSITE: *Xenograpsus testudinatus* lives around shallow hydrothermal vents off Taiwan, surviving on zooplankton that rains down from the surface – killed by the hot sulphurous plumes.

CRABS AS SCAVENGERS

Crabs can be voracious scavengers, and most species will take the opportunity to feed if they find something dead (see, for example, the test carried out by expeditioners searching for the remains of Amelia Earhart, page 45).

A fascinating example of the opportunistic nature of foraging is shown by the hydrothermal vent crab (*Xenograpsus testudinatus*), which lives in very high densities around shallow-water, sulphur-rich hydrothermal vents found off Taiwan. Researchers were initially mystified by how so many crabs could live in such an acidic environment so low in nutrients, but then observed that vast numbers were only swarming out of crevices at the top and bottom of the tides. It is only then, when the currents stop flowing, that the zooplankton on which the crabs depend rains down from the surface like marine 'snow' – killed by the hot sulphurous plumes from the vent (see also page 117).

VEGETARIANS

Vegetable matter is predominant in the diet of many brachyurans. Some species have claws specially designed to scrape algae from rocks, snip off fronds of larger macro-algae, or to cut and shred leaves and seedlings in mangrove and terrestrial forests. Plant structural carbohydrates, cellulose and hemicellulose, are a major source of potential energy, but are difficult to digest without the enzyme cellulase needed for their breakdown. Herbivorous animals such as cows, horses and sheep, and even some insects, have a culture of symbiotic bacteria in their intestinal tract that produces the necessary cellulase enzymes, but no such bacteria have ever been detected in crabs. It was speculated that some mangrove crabs took leaves and plant material into their burrows to allow time for bacterial and fungal colonies to do the work of breaking them down. However, recent research has shown that the digestive gland of gecarcinid land crabs is able to manufacture the specialized enzymes needed to metabolize cellulose and hemicellulose. No doubt further research will show that this ability has evolved across many crab lineages.

CRABS AS PREY

Crabs are hunted and eaten by a wide range of hungry predators – virtually anything that has worked out how to avoid their claws and has the power to crack through their shells. Marine and estuarine crabs are targeted by octopus and squid, bottom-feeding fish and eels, crocodiles and alligators, and even seals and otters. Intertidal and freshwater crabs are a favourite food of waders and other seabirds, owls, monitor lizards and specialist crab-eating snakes. A variety of mammals even join in on crab feasts when they can, particularly rats, raccoons and the crab-eating macaque monkey of Southeast Asia. Strangely enough, the crab-eater seal does not feed on crabs at all – it is a specialist predator of Antarctic krill.

OPPOSITE: 1. A crab plover on a beach in Thailand has removed the claws from a green swimming crab (*Thalamita* species), to make it easier to manage; 2. A European otter with a velvet crab (*Necora puber*) on the Isle of Mull in Scotland; 3. An American alligator with a large blue crab (*Callinectes sapidus*).

RIGHT: In a flat world, the evolution of long, vertical eyestalks (with an elongated cornea wrapped around the top) has allowed fiddler crabs and their relatives to divide their field of vision into an upper hemisphere for spotting predators, and a lower hemisphere for viewing other crabs, based on the horizon line. On a sloping beach, the eyes always will tilt vertically to align with the horizon.

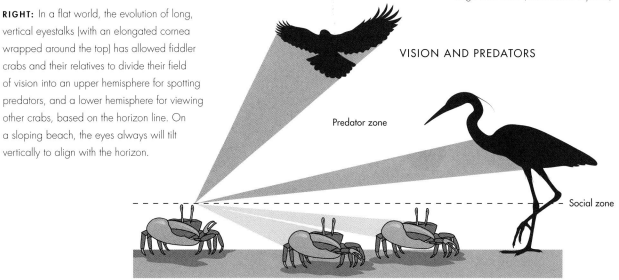

A MOST SUCCESSFUL GROUP... 109

Shallow subtidal, intertidal, semi-terrestrial and terrestrial species can have many enemies. For example, the east Australian soldier crab (*Mictyris longicarpus*) is attacked by a variety of predators: at least three species of birds (strawnecked ibis, mangrove kingfisher and eastern great egret); toadfish; two other crab species (the ghost crab *Ocypode ceratophthalmus*, and a grapsid, *Metopograpsus frontalis*); and a sand moon snail (*Conuber sordidus*), which predates on juvenile crabs by enveloping them with its foot before drilling through the carapace.

Crabs are also the dietary mainstay of the chain moray eel (*Echidna catenata*). Found along the rocky shores of the tropical western Atlantic, from Florida to Brazil, this eel has blunt teeth specially designed to crunch crab shells. It seems particularly fond of the Sally Lightfoot shore crab (*Grapsus grapsus*), a fast and agile species with an almost unique ability to leap high in the air in order to reach adjacent rocks. The eel hunts on the falling tide when the crabs are migrating down the shore, and around the tide pools where they are feeding on newly exposed algae. If necessary, the eels can launch themselves fully out of the water, chasing the crabs up to 5 m (16 ft) across the rocks. A special mucus coating helps protect morays from injury as they traverse sharp barnacles and oyster shells. If a crab is lucky, it is fast enough to prance away; unlucky ones are torn apart, or swallowed whole.

LEFT: An American white ibis (*Eudocimus albus*) eating a crab. The long, curved beaks of this and other wading birds are perfect for probing the mud in search of food.

BELOW LEFT AND RIGHT: Sally Lightfoot crabs (see page 130) inhabit tropical rocky shores of the East Pacific and Atlantic. In the Americas they are hunted mercilessly by the crab specialist, the chain moray eel.

LEFT: The Atlantic ocellate box crab (*Calappa ocellalta*) is often followed closely by fish, like this flounder. But rather than attack, the fish are only intent on snapping up any small invertebrates disturbed by the crabs while they search the bottom for their own food.

BELOW: A white-bellied mangrove snake (*Fordonia leucobalia*; Homolopsidae) eating a crab in the swamps of northern Australia. Three species of this family specialize in eating brachyurans.

Amongst reptiles, at least three species of snakes are specialized crab predators. *Fordonia leucobalia* and *Gerarda prevostiana* are water snakes that share the brackish mangrove swamps of the Indo-Malaysian region. *Fordonia leucobalia* attacks small hard-shelled crabs less than half the size it could potentially swallow, seemingly intimidated by the claws and spines of larger animals. Unlike the typical open-mouthed strike of other snakes, *Fordonia* strikes with its 'chin' in order to first pin the crab to the ground (a bit like putting a foot on it!). It then coils its body around the wriggling crab and swallows it whole. *Gerarda prevostiana* (Gerard's water snake) specializes in eating large crabs that are newly moulted and still soft, ripping off bite-sized chunks. However, because the crab shells harden quickly, the snake has a narrow window of culinary opportunity. Smith's mountain keelback (*Opisthotropis spenceri*), although an unrelated snake, has evolved a similar method of crab attack. Endemic to the fast-flowing mountain streams of northern Thailand, it seeks out newly moulted freshwater crabs. Holding the victim in its coils, it removes and swallows the claws and legs one by one, then opens the body to feed on what remains.

Small shore mammals also relish brachyurans and other intertidal crustaceans. Small mangrove crabs are an important component of the diet of the endangered tropical Australian native water mouse (*Xeromys myoides*), which inhabits mangrove forests, saline grasslands and freshwater swamps. This nocturnal predator attacks methodically, first biting off the eyes, then tearing off the claws, and finally extracting flesh from the crab's carapace by cutting through the thinner under-surface with its teeth.

A MOST SUCCESSFUL GROUP...

CRAB ENVIRONMENTS

CRABS CAN BE FOUND in near-freezing abyssal ocean depths, scorching intertidal beaches and swamps, arid deserts and tropical montane forests. Given that crabs are very abundant in some environments, they can play an important role in the function of both aquatic and terrestrial ecosystems.

MUD AND SAND
Mangrove swamps, and their associated muddy-sand flats, are very harsh places to live. Animals here must deal with the daily flood and ebb of the tides, often very high temperatures, widely fluctuating salinities and pH, anaerobic mud, and constant pressure from a wide range of predators. Nevertheless, such swamps can be extremely rich in invertebrate life, with crabs often making up to five times the biomass of all the other invertebrates combined.

Active burrowing by vast numbers of crabs displaces and mixes a large part of the top 20 to 30 cm (1 ft) of mud every year, thus enriching it by bringing organic material back up to the surface. Burrows also have a vital function in aerating the mangrove soil, which due to its finely particulate muds, can quickly become anaerobic. Many of the larger crab species burrow between the trunk buttresses, or amidst the tangle of prop roots. While this is done to avoid predators, it allows atmospheric oxygen and fresh tidal water down amongst the root systems, assisting in organic decomposition and encouraging forest growth.

Herbivorous crabs, especially sesarmids, eat and bury enormous quantities of leaves, fruit and seedlings, promoting the action of bacteria and fungi, and the chemical processes of decay. This ensures that nutrients present in the vegetation are released back into the mangrove ecosystem, forming the basis of the complex food chain that culminates in the vast numbers of large coastal fish, prawns and crabs that are dependent on these habitats.

Activity patterns must be tuned to both daily and tidal rhythms. Lower and mid-shore crabs are typically diurnal and have only a few short hours for feeding and social activity. With increasing height up the shore, temperature and salinity variation can become extreme and water becomes a precious resource, so many mangrove and salt-marsh crabs stay in their burrows and only emerge at night, or in the early morning or evening.

Long periods out of water also puts severe strain on gills that evolved to work underwater. Therefore, many mangrove crabs have reduced the size of the gills and developed 'pseudo-lungs' – highly vascular tissue that can absorb atmospheric oxygen. The mangrove specialist crabs of the family Sesarmidae have the remarkable ability to recycle and re-oxygenate water by pumping it in a fine film across the front and sides of their bodies (see page 80).

OPPOSITE: Mainly surviving on mangrove leaves, this Australian mangrove crab (*Neosarmatium australiense*) is meeting its protein needs with a 'flame fiddler' that strayed too close.

BELOW: Crabs form the major component of the animal biomass, and through their feeding and burrowing activities play a key role in the healthy functioning of the mangrove ecosystem.

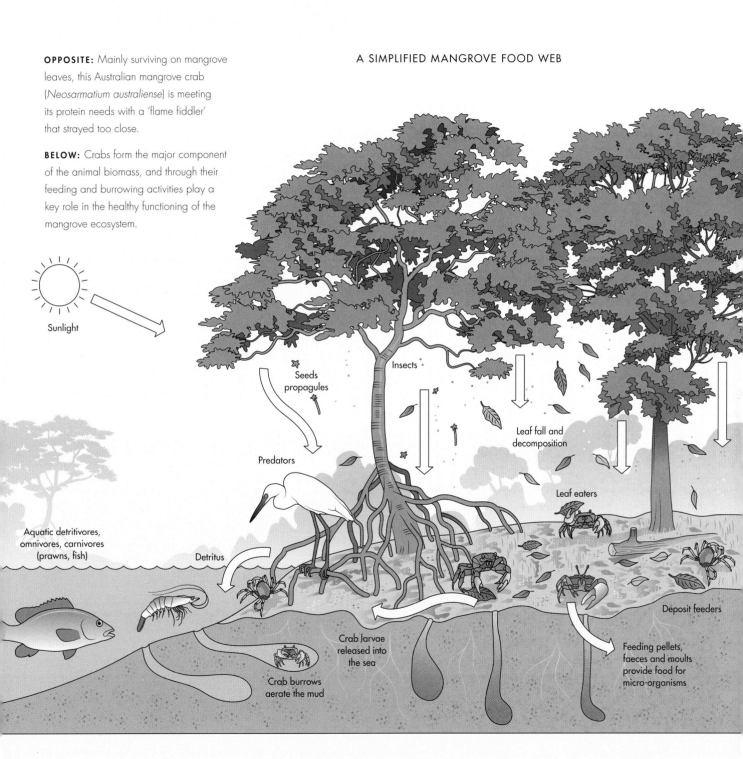

A SIMPLIFIED MANGROVE FOOD WEB

CRAB ENVIRONMENTS 113

ROCKY SHORES

Intertidal rocky shores can be difficult environments for crabs, especially where shores are exposed to wave action. Crabs have only a narrow window of activity between the tides to feed, and this can be shortened by the intense heat as the rocks start to bake under a hot sun. Shore crabs are usually forced to take shelter under rocks and ledges, or in narrow cracks and fissures. Tidal pools can be refuges if they are low on the shore and have some water exchange, but trapped pools can become warm and hypersaline, so need to be avoided.

Crabs on rocky shores are also exposed to attack from the air by hungry birds, so they are, by necessity, swift movers and can vanish in the blink of an eye. Typically, their legs are tipped with strong, sharp claws that give them purchase on uneven rocky surfaces, and strong spines on the legs can be used to lock them into their narrow hiding places, despite the efforts of an eager predator, or the rush of a wave breaking and receding over their heads.

Many shore crabs are herbivorous, scraping algae from the rock surfaces exposed by the falling tide. The most conspicuous of these are the Sally Lightfoot crabs and their relatives (family Grapsidae), although closer to the water some xanthoid reef crabs, such as species of *Eriphia*, can also be present.

TOP RIGHT: The Indo-West Pacific red-eyed reef crab (*Eriphia sebana*) inhabits tropical rocky shores. Active by night, it hides in crevices during the day. Powerful claws are perfect for crunching snail shells.

RIGHT: The slender rock crab (*Grapsus tenuicrustatus*) mainly lives by scraping algae from the rock surfaces. Its thin body slides easily into crevices, while its broad legs lock it in place.

CORAL REEFS

The popular concept of a coral reef conjures up images of colourful branching staghorn corals and large schools of vibrant fish, and much of the research conducted into the ecology of reefs has concentrated on these two iconic groups. But reefs are also home to a myriad of other invertebrates, and the diversity and importance of crabs in reef ecology has been largely overlooked. There is now growing evidence that algal grazing by legions of brachyurans, both small and large, can be just as important in controlling reef community structure and health as grazing by fish. The black-fingered xanthoid reef crabs are particularly abundant and important; field experiments have demonstrated that three species, *Chlorodiella nigra*, *Cyclodius ungulatus* and *Leptodius exaratus*, remove significant quantities of epibiotic algae from dead coral branches on Australia's Great Barrier Reef. In doing so, these crabs play a vital role in maintaining spaces within the branching corals. This in turn supports fish recruitment, provides fishes with nocturnal and diurnal refuges from predation, and enhances habitat complexity and biodiversity.

Literally thousands of crab species are co-dependent on coral reefs, with some forming a commensal relationship with the corals themselves. For example, the 'guard' crabs in the families Trapeziidae and Tetraliidae always live within the branches of coral, where they are protected from predation, but they also protect their coral host, and will attack any coral feeders that try to nibble on polyps – even crown-of-thorns starfish (see page 121)!

BELOW LEFT: Common in coral reef environments, the red spider crab (*Schizophrys aspera*) tries to disguise its presence by growing encrusting organisms on its shell.

BELOW: A batwing crab (or queen crab; *Carpilius corallinus*), native to the coral reefs of the Caribbean and Bahamas. It is a large species, growing to about 12 cm (5 in).

CRAB ENVIRONMENTS

THE DEEP SEA

The 'deep sea' starts at around 200 m (650 ft) depth, or the point where the continental shelf becomes the continental slope. For crabs, this marks the beginning of a noticeable shift in species composition, with most shallow-water species disappearing, and a new suite of deep-water inhabitants taking their place. Much of the ocean floor consists of relatively sparsely populated, vast plains of soft muddy sediments, but the marine life around features such as seamounts, hydrothermal vents, ridges and trenches is often both rich and abundant.

Many deep-water species, especially those living below 500 m (1,640 ft), exhibit what are termed 'K-selected' life histories (extreme longevity, late age of maturity, slow growth and low reproduction rates). This means they are inherently vulnerable to over-exploitation. Bottom trawls can quickly destroy a complex habitat (especially those on seamounts) that has potentially been built over millennia by slow-growing invertebrates. Significant irreparable damage has already been done to a number of areas around the globe, and some deep-water crab fisheries have already collapsed because of imprudent over-exploitation (see Chapter 5).

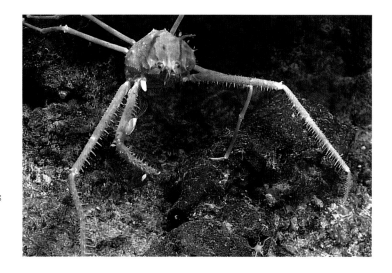

Crabs, like so many other animals, have evolved some fascinating adaptations to living in one of the most hostile environments on the planet. Many are pink or red in colour because red light wavelengths do not penetrate into deep-water, so red actually appears black, and makes them less visible to predators and prey. Crabs of the Ethusidae are all specialist deep-water species, with one genus *Ethusina* being found at the greatest depths and completely blind; the aptly named *Ethusina abyssicola* has been found more than 4,000 m (13,000 ft) down. The greatest diversity of deep-water crabs, however, is typically much shallower – from about 200–800 m (650–2,600 ft). In this zone there are many strange spider crabs such as *Cyrtomaia* species that stride across the bottom on long, spindly legs.

MARINE ENVIRONMENTS

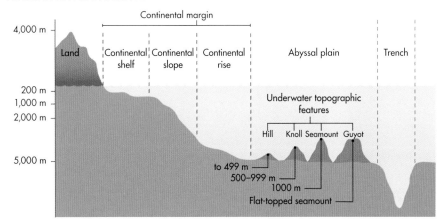

ABOVE: Deep-sea spider crabs such as species of *Cyrtomaia* (family Inachidae) are believed to catch small fish by spearing and holding them with the spines that line the inside of their long front legs and claws.

LEFT: A schematic representation of different marine environments, from the intertidal to the abyssal depths. The deepest part of the ocean is the Challenger Deep trench (over 11 km [36,200 ft] deep) in the western Pacific, which runs for several hundred kilometres.

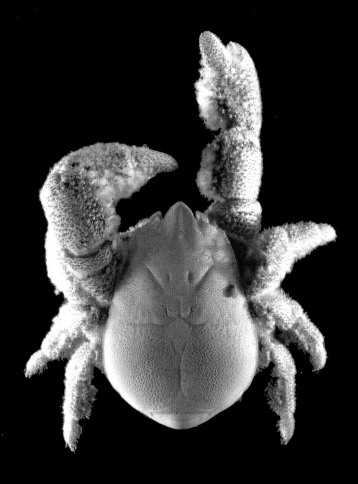

LEFT: The deep-sea Hoff crab (*Kiwa tyleri*) lives around hydrothermal vents. Like the Yeti crab, it feeds on sulphur bacteria that it 'farms' on the soft setae of its arms and 'chest'.

HYDROTHERMAL VENTS

In some special places, water drawn into the sea floor is superheated as it passes through hot sub-surface rocks, and is then expelled through fissures as plumes of hot fluid, rich in sulphides and other inorganic compounds. These phenomena are called hydrothermal vents and are commonly found near areas of deep-sea volcanic activity. They were only discovered in 1976, by scientists from the Scripps Institution of Oceanography. Expeditions using submersibles and ROVs quickly discovered not only a new geological phenomenon, but also a rich variety of new species, at densities 10,000 to 100,000 times greater than that of the surrounding seabed. Too deep for the light of day to penetrate, there can be no conventional solar-powered photosynthesis to generate energy and growth. Instead the food chain is based on thick mats of chemo-synthetic bacteria that use methane and sulphur compounds in the fluids emanating from the vents to synthesize organic matter. These diverse bacteria – over 250 separate strains have been identified – thus form the basis of a unique food web that can support creatures found in no other ecosystem.

One of the early animal discoveries was an amazing new family of crabs, the Bythograeidae. Sixteen species have since been described in six different genera, and each new vent area explored seems to have another new species. Bythograeids are remarkably adapted to their extreme lifestyle. They are blind in the conventional sense, but able to perceive light in the darkness with eyes that detect portions of the spectrum produced by hydrothermal vent chemistry (see page 129). Bythograeids have also developed the ability to detoxify themselves of the poisonous sulphides that are so characteristic of vent fluids. The digestive gland of the mid-gut secretes sulfite oxidase, which converts sulphides into harmless thiosulfate that can then be excreted.

Another famous vent inhabitant, the blind Yeti crab (*Kiwa hirsuta*), excited much interest when it was first discovered in 2005 at 2,200 m (7,200 ft) depth on the Pacific–Antarctic Ridge. Although popularly called a crab, it is actually a type of squat lobster. There are now four known species of *Kiwa* and all live only around hydrothermal vents. They are remarkable because they actively farm sulphur bacteria on the soft setae of their claws, legs and bodies. Some species actually wave their furry claws close to the hot sulphurous plumes to encourage bacterial growth! The bacteria are harvested and eaten using specially modified mouthparts.

LEFT: The Malawi blue crab (*Potamonautes orbitospinus*) endemic to Lake Malawi, in East Africa, plays an important role in its freshwater ecology.

OPPOSITE: One of a group of 'vampire crabs' dependent on pitcher plants, this tiny *Geosesarma* species from Sumatra raids plants for insects trapped in the bottom.

FRESHWATER AND TERRESTRIAL CRABS

True freshwater crabs belong to three families: Gecarcinucidae, Potamonidae and Trichodactylidae (the last only being found in Central and South America). Freshwater crabs have been enormously successful, comprising about 20 per cent of the over 7,000 species of crabs so far discovered across 100 families. Unlike all other crabs, they are no longer tied to the sea for breeding because their larval development takes place completely within the egg, and they give birth to baby crablets. They are restricted to warmer tropical and subtropical regions worldwide, but can be found from lowland forests up into high rugged mountains and even into deserts, wherever there are rivers, streams, waterfalls or wetlands.

Freshwater crabs are typically amphibious to varying degrees, and a few can be almost fully terrestrial – but access to at least some fresh water is always necessary, even if it is only seasonal. As mentioned in Chapter 2, the Australian inland freshwater crab (*Austrothelphusa transversa*) can survive in desert conditions through droughts of six years or more. It does this by plugging its burrow with earth and going into a dormant state, staying just alive by using stores of fat in its tissues.

In some areas freshwater crabs can be of major ecological importance because they are relatively large animals that can often occur in great numbers. For example, research on the streams and dams of the East Usambara Mountains of Tanzania has estimated that they comprise around 90 per cent of the total invertebrate biomass. Being omnivorous, they take advantage of whatever food is available, including fresh or fallen leaves, seeds and other vegetable matter, aquatic insects, snails, dead frogs, snakes, or other carrion. Some are even active predators of small vertebrates. However, their role as detritus-shredders makes them of great importance to nutrient recycling and food web structure. When the North American red swamp crayfish (*Procambarus clarkii*) was introduced to Western Kenya, this aggressive alien outcompeted and replaced the native crabs, leading to an unforeseen decline in the clawless otters that had previously relied on a ready supply of crabs for food. Freshwater crabs are similarly popular with a wide variety of predators throughout the world, including humans.

Some crabs can take advantage of even the smallest amounts of fresh water trapped in plants (phytotelmata). Water can form tiny reservoirs in cavities in trees, or in the case of tropical bromeliads, it accumulates in the cups formed where the leaves meet in a rosette. In coastal Pacific mangrove forests, several species of tree-climbing sesarmid crabs use bromeliad phytotelmata for concealment, as a source of food, and for breeding. In terrestrial rainforests, particularly through Africa and Madagascar, there are also numerous freshwater crabs that rely on phytotelmata as a source of water for breeding. A particularly interesting form of phytothelma is the water captured in the bottom of carnivorous pitcher plants (*Nepenthes*). These 'pitchers'

can be home to a range of aquatic insects, especially mosquito larvae, and some species of crabs in the genus *Geosesarma* (Sesarmidae) are specialized to invade the plants to feed on the insects and larvae contained within.

One family of large terrestrial crabs, the Gecarcinidae, has been very successful throughout the tropical regions of the world, and sometimes occur in very large numbers. Despite living much of their lives out of the water, they are considered semi-terrestrial because they always need to migrate back to the sea to release their eggs and complete their life cycle. These crabs are featured more extensively in Chapter 4.

CAVES

In many ways caves are like islands, their own small universes cut off from the rest of the world. And like islands, many caves have been isolated for such a long time that they have evolved their own strange endemic specialist species called 'troglobites'. Troglobitic crabs typically have adaptations that help them live in these permanently pitch-black environments, including the loss of functional eyes and shell pigmentation, and long, thin walking legs. Caves have been invaded by all the true freshwater families, but some genera of more typically marine crab famiIes (such as Goneplacidae, Hymenosomatidae and Varunidae) have also become specialized troglobites, especially in caves that have a connection to the ocean ('anchialine' caves).

One particularly interesting discovery was a blind freshwater crab species called *Trogloplax joliveti*. Only described in 1986, this crab was exciting because it was unrelated to any other freshwater crabs known at that time. In fact, it was so unusual that it was placed into its own subfamily (Trogloplacinae) within the Goneplacidae. It is still only known from a single karst cave system at 650 m (2,100 ft) altitude on the island of New Britain, Papua New Guinea. In 1989, its closest relative, *Australocarcinus riparius*, was described from northern Queensland, Australia. That species lives in freshwater montane rainforests, but is otherwise a small nondescript crab that shows none of the remarkable cave adaptations of *Trogloplax*. It is, however, from such an ancestral crab entering deeper and deeper into the gloom of caves that *Trogloplax* must have evolved.

RIGHT: *Trogloplax joliveti* is a blind freshwater troglobitic species described in 1986 from a single karst cave system on the island of New Britain, Papua New Guinea.

CRAB ENVIRONMENTS

LIVING WITH OTHERS: SYMBIOSES

NOT ALL CRABS LEAD FREE-LIVING EXISTENCES. Many are dependent to various degrees on other animals, and often that need is reciprocated (*mutualism*). In fact, six whole families of crabs fall into the category of *obligate symbionts* (species that cannot survive without their host species), and some other families also have species that have adopted the same way of life.

Pea crabs (families Pinnotheridae and Aphanodactylidae) are symbiotic crabs that may be familiar to many, as they can often turn up at the dinner table as a surprise extra inside the shells of mussels. Most pinnotherids live in the mantle cavity of bivalve molluscs (mussels, clams and their ilk), living off food particles from their hosts' gills. Pea crabs are less commonly associated with sea squirts (ascidians) and echinoderms (especially holothurians, or sea cucumbers). One species of pea crab, *Pinnixa tumida*, enters through the anus of its host, *Paracaudina chilensis*, to live on the faecal matter and mucus in the sea cucumber's gut! Other pinnotherids are specialized to share the burrows and tubes of bristle worms (polychaetes), mud lobsters, peanut worms (sipunculans) or spoon worms (echiurans).

Crabs from the families Cryptochiridae, Domeciidae, Tetraliidae and Trapeziidae are all found predominantly on stony corals, but sometimes also on other colonial cnidarians such as fan corals (gorgonians) and black corals (antipatharians). Most coral-inhabiting crabs tend to eat coral tissue, and the bacteria-rich mucus that corals produce to keep their surfaces free of sediment and detritus. In return for protection and a ready food source, species of *Trapezia* (family Trapeziidae) use their sharp claws to protect their coral patch from predators. Others, like species of *Cymo* (family Xanthidae) are more timid, but wonderfully adapted to wrap their body around the coral branches.

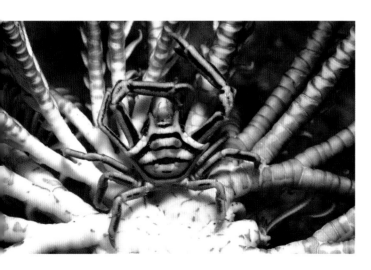

LEFT: The horned crinoid crab (*Ceratocarcinus longimanus*) belongs to an Indo-West Pacific subfamily of Pilumnidae reef crabs. They all live associated with echinoderms, but shapes and colours vary according to the host.

ABOVE: A pea crab (*Pinnotheres pisum*) inside a mussel. Not always a harmless guest, it can inhibit egg production and cause gill damage and weight loss in the host.

Gall crabs (family Cryptochiridae) are particularly strange. The female marsupial crab (*Hapalocarcinus marsupialis*) lives on the tips of branching corals, irritating it such that it slowly grows around the annoying crab. The resultant calcareous cage provides the crab a protective home – but one that she can never leave. There is only enough room between the coral 'bars' for the tiny dwarf male to enter for mating, and for the eggs to be released. Other cryptochirids typically form holes or depressions ('pits') in which to live in their chosen coral hosts, which they often match in appearance (see page 57).

A completely different type of symbiosis exists between a group of small, colourful xanthid crabs and sea anemones. Species of *Lybia* (and some related genera) carry two tiny sea anemones in their claws. Food particles that the anemone gathers in its tentacles are stolen by the crab, and the stinging anemones may also be brandished to fend off small predators. If an anemone is lost, the crab will replace it by cutting the remaining one in half. Each half will then grow into new full-size anemones – primitive but effective cloning. In fact, some may argue that this relationship is more like tool use than symbiosis (see page 86).

BELOW: A honeycomb guard crab (*Trapezia septata*) mounts an aggressive response by nipping at the feet of a crown-of-thorns starfish threatening to eat its coral home.

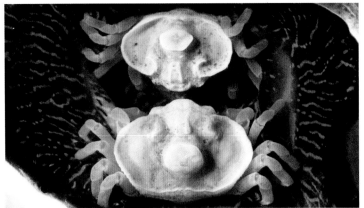

Echinoderms (particularly sea urchins, sea cucumbers and feather stars) also have their share of crab gatecrashers. For example, 'eumedonine' crabs in the family Pilumnidae are echinoderm specialists. The pentagonal sea urchin crab (*Echinoecus pentagonus*) lives underneath urchins, between the spines, with adult females even setting up house inside the urchin's rectum. A related crab, *Hapalonotus pinnotheroides*, looks like a very large, round, smooth pea crab, and lives inside sea cucumbers in a membranous cyst formed near the cloaca. Here it spends its existence filtering food from the water passed through the holothurian's respiratory tree. Other eumedonines come in a wide and weird variety of shapes and colourful patterns, depending on their echinoid host. They typically feed on tube feet and skin tissue, but those on crinoids live off the rich mucus that the feather stars make to transport their own food along their arms to their mouth.

Symbiotic relationships can even be found amongst species traditionally considered free-living. For example, some Indo-West Pacific swimming crabs (Portunidae) depend on sea cucumbers. *Lissocarcinus orbicularis* lives around a sea cucumber's oral tentacles, and moves in and out of its intestine through the anus. And a related species, *Lissocarcinus laevis,* often spends time in the company of sea anemones (see page 136).

LEFT: Female pentagonal sea urchin crabs living in the rectum of sea urchins eventually become trapped in a calcified gall, where they are visited by the much smaller males.

ABOVE: *Xanthasia murigera* is a specialized member of the pea crab family (Pinnotheridae). Males and females live in pairs amongst the mantle folds of giant reef clams.

CHRISTMAS ISLAND BLIND CAVE CRAB
Christmaplax mirabilis

FAMILY:	Christmaplacidae
DISTRIBUTION:	Indigenous to Christmas Island
HABITAT:	Saltwater limestone caves
FEEDING HABITS:	Predator, scavenger, omnivore; small gastropod snails probably form an important part of its diet
NOTES:	*Mirabilis* is Latin for 'amazing' or 'remarkable', expressing how scientists felt upon first finding it
SIZE:	To about 12 mm (1/2 in) across the carapace

THIS REMARKABLE CRAB WAS first described only in 2014, and represented a new genus and new species, but was also so different that it needed to be placed in a whole new family, Christmaplacidae. It lives in pitch-black underwater caves that intrude into the ancient limestone reefs that surround Christmas Island. In 2017, a second genus and species, *Harryplax severus*, was added to the family – also notable for living deep within the interstices of subtidal coral rubble and rocks. While it too has reduced eyes, it can still functionally use them.

WORLDS APART

Limestone karst landscapes occur around the world, with extensive cave systems forming over thousands or even millions of years by subterranean water gradually dissolving huge cavities in the remains of long-buried coral reefs. These caves are like time capsules that hold the evidence of the geological history and past climates of the region. They also act as isolated landlocked 'islands' in which unique cave specialist 'troglobite' species have evolved. Troglobitic crabs share a number of similar convergent adaptations to help them survive, including the loss of functional eyes and shell pigmentation (giving them a ghostly look), and long, slender walking legs. This is why they appear to be so different from more familiar-looking relatives living nearby in sunlit waters or on land – they truly live in a world apart.

KEPT IN THE DARK
A life in blackness means eyes are no longer needed; colour is not important so the crabs become pale; and long, thin walking legs are useful for probing and sensing the world around them.

GAIMARD'S SPIDER CRAB
Leptomithrax gaimardii

GIANT SPIDER CRABS MOSTLY LIVE a relatively solitary life, widely dispersed across the sea floor in the deep, dark, cold waters of southern Australia. Little is known of their general biology, but once a year, during winter, they leave behind their normal lives to gather in their hundreds, or thousands, at various locations in southern Victoria, Tasmania and South Australia. Crabs emerge into the shallows, slowly grouping together in small clusters during late summer to early autumn, before finally gathering en masse for the big winter event (see page 146). They often return to the same locality for a number of years in a row, before suddenly inexplicably moving on to a new festival site.

STRONGER TOGETHER

These huge aggregations were only discovered in recent years, but this phenomenon is now a popular annual event, enthusiastically monitored by divers and photographers. Such behaviour is not unique to *Leptomithrax gaimardii* but has also been observed by related species of majid spider crabs, in particular the common European spider crab (*Maja brachydactyla*). The prevailing theory is that the crabs come together to protect each other during mass synchronized moulting, when they are most vulnerable to large fish, rays and octopus. Hard-shelled crabs tend to stay around the perimeter, to better protect the newly moulted and vulnerable soft-shelled crabs towards the centre. In *Maja brachydactyla*, there is also accompanying mating activity, though this has not yet been directly observed for Gaimard's spider crab.

FAMILY:	Majidae
OTHER NAMES:	Giant or great spider crab
DISTRIBUTION:	Endemic to temperate southern Australia
HABITAT:	Reef and sand areas; to 820 m (2,690 ft) depth, but typically much shallower
FEEDING HABITS:	Omnivorous scavengers
NOTES:	Newly moulted crabs are bright orange, but gradually fade to brown; macro-algae and sponges often grow on the shell
SIZE:	To about 16 cm (6¼ in) carapace width, with leg span to 40 cm (15¾ in)

A LITTLE PRIVACY PLEASE!

Since the phenomenal mass migration of these large crabs was discovered in the shallow waters near Melbourne, Australia, it has become an increasingly popular annual pilgrimage for ecotourists, divers and recreational fishers to witness this amazing event.

HOURDEZI'S HYDROTHERMAL VENT CRAB
Austinograea hourdezi

FAMILY:	Bythograeidae
DISTRIBUTION:	Southwest Pacific: Manus Basin, North Fiji Basin and Lau/Tonga region
HABITAT:	Associated with the sides of hydrothermal vents; 1,600–2,630 m (5,250–8,630 ft)
FEEDING HABITS:	Predator and scavenger; grazes on bacterial mats
NOTES:	Bythograeids have an extraordinary ability to detoxify poisonous sulphides – their mid-gut digestive gland secretes sulphite oxidase which converts sulphides into harmless thiosulfate that can then be excreted
SIZE:	To about 35 mm (1 3/8 in) carapace width

AS PROOF OF THEIR VERSATILITY, crabs have managed to conquer what must be one of the harshest and most alien environments on the planet – hydrothermal vents. At depths reaching 2.7 km below the surface, pressures of 250 atmospheres, and in the shadow of plumes of superheated sulphurous water pumped from below the Earth's crust through volcanic chimneys, things do not get much tougher.

SEEING IN THE DARK

The eyes of hydrothermal vent crabs such as *Austinograea* are unique within the Brachyura. In the pitch darkness, normal vision is impossible, and like crabs that live in caves, their eyes are substantially atrophied, with the eyestalk fixed rigidly in the orbital socket. Research shows that the crab's planktonic zoeae larvae – which develop in the mesopelagic (or twilight) zone, closer to the surface – have essentially normal crab eyes, with visual pigment sensitive to the blue light spectrum of the water depth in which they occur. By the time the crabs have become adult and settled around the vents, however, they have lost all image-forming capability, and instead have developed unique, highly sensitive 'naked-retina' eyes, better able to detect light produced by hydrothermal vent chemistry in the near-infrared portions of the spectrum.

MEMBER OF A SPECIALIST COMMUNITY
A fascinating community of animals is specialized to live around hydrothermal vents. Here *Austinograea hourdezi* is shown walking amongst mussels (*Bathymodiolus brevior*), gastropods (*Ifremeria nautilei*) and a sea anemone (probably *Cyananthea hourdezi*).

SALLY LIGHTFOOT CRAB
Grapsus grapsus

THE STUNNINGLY HANDSOME Sally Lightfoot crab has captured human attention for longer than most crabs. It was first described, as *Cancer grapsus*, by Carl Linnaeus in 1758, and later collected by Charles Darwin on his visit to the Galápagos Islands in 1835. The origins of the common name are unclear – the most likely explanation is that it originated in Jamaica, in honour of a local dancer who was no doubt both colourful and light on her feet! Indeed, *Grapsus grapsus* is famous for its ability to leap into the air, dancing from rock to rock.

UNFUSSY EATERS

These crabs live intertidally in the harsh environment of wave-crashed rocky shores. Younger crabs are typically found closer to the water, with larger crabs dominating above the spray line. They are omnivorous, eating a wide range of foods, and while algae is on the menu, they can also be aggressive predators, eating everything from turtle hatchlings to the eggs and chicks of nesting shore birds, barnacles, mussels, and even each other. The presence of numerous crabs missing legs and claws attests to narrow escapes from their aggressive compatriots; in Peru, a large female was even observed to cannibalize a smaller male directly after mating! A study from the Galápagos found that they will even eat ticks from the skin of the large marine iguanas with which they share the rocks.

FAMILY:	Grapsidae
OTHER NAMEs:	Red rock crab
DISTRIBUTION:	Pacific coast of Mexico, Central America, South America (as far south as northern Peru), and nearby islands, including the Galápagos; also along the Atlantic coast of South America
HABITAT:	Exposed oceanic rocky shores; intertidal
FEEDING HABITS:	Omnivore and scavenger
NOTES:	Few attackers are fast enough to catch a Sally Lightfoot, but the chain moray eel is a specialist, with blunt teeth that can crunch through crab shells
SIZE:	To about 87 mm (3 3/8 in) carapace width

MASTERS OF THEIR DOMAIN

Crabs transform from well-camouflaged green-grey youngsters to become predominantly red, orange and sky blue as sexually mature adults, their colours glowing against the dark volcanic rocks upon which many live. By that stage, few natural predators are big or fast enough to challenge their rock-face supremacy.

RED-KNEED SOLDIER CRAB
Mictyris longicarpus

FAMILY:	Mictyridae
OTHER NAMES:	East Australian soldier crab
DISTRIBUTION:	Indigenous to eastern Australia, but a number of other species occur through the central Indo-West Pacific region
HABITAT:	Intertidal muddy-sand flats
FEEDING HABITS:	Deposit feeder
NOTES:	The tumbling multicoloured herd distracts predators and helps provide safety in numbers.
SIZE:	Over 25 mm (1 in) across

MASSIVE ARMIES OF SKY-BLUE, cream and maroon soldier crabs scrambling across the muddy sand at low tide are a fascinating sight. Soldier crabs must fit all their active living into the low tide period. To do this they rely on directly breathing air via sophisticated 'lungs' (branching finger-like airways that line the inside of the globular carapace). They use their gills so little that they even seal off an air-filled subterranean chamber prior to the incoming tide.

SAND SIFTERS

Their diet consists almost entirely of plant detritus, microscopic algae and meiofauna (minute animals living between the sand grains). Remarkably, despite sieving through vast quantities of muddy sand, no sand grains pass into the stomach. This is achieved by creating a 'flotation chamber' inside their large, bulbous outer mouthparts. Lighter food items are then scooped from the sandy slurry by specially modified setae (hairs) that act as strainers. The heavier sand grains are then rejected. Soldier crabs are also famous for having mastered the ability to walk forwards and diagonally as well as the traditional sideways walk for which most crabs are so well-known.

UNIQUE BURROWING STYLE

The en-masse wandering habit of these small crabs makes building permanent burrows out of the question. But any disturbance from a predator, or other beach visitor, and the whole army will disappear in a matter of seconds by corkscrewing into the wet muddy sand.

LEWINSOHN'S SPONGE CRAB
Lewindromia unidentata

'SPONGE CRAB' IS THE NAME typically applied to the large family Dromiidae, which consists of mostly small species that only a few people who know what to look for will ever see. Why? Because they are masters of disguise. These little chaps typically hide under a tight-fitting cap of living sponge that they hold with their last two pairs of legs. These specialized legs are no longer used for walking, but are positioned over the back of the carapace, and each is armed with spinous claws that hold the crab's chosen covering firmly in place.

FASHIONING A SHELTER

Crabs select a piece of sponge of about the right size, then use their front claws to hollow it out so that it fits over their shell snugly. Although sponge is commonly used – probably because it is easy to sculpt – many dromiids such as *Lewindromia unidentata* will use a wide range of other organisms, including soft corals, anemones, and compound or solitary ascidians (sea squirts). Once covered, the crab is almost impossible to detect unless it moves. Not only does the sponge provide wonderful camouflage, but it and the other invertebrates that are used are typically quite toxic, so would-be crab predators are deterred from investigating.

FAMILY:	Dromiidae
OTHER NAMES:	Red-spot sponge crab
DISTRIBUTION:	Indo-West Pacific: East Africa to Hawaii, north to Japan, south to Australia, Kermadec Islands and Easter Island
HABITAT:	Rocky and coral reefs; subtidal to about 100 m (330 ft) depth
FEEDING HABITS:	Omnivore; grazer, scavenger
NOTES:	Mostly active nocturnally for even better protection
SIZE:	To about 34 mm (1 3/8 in) carapace width

GROWING UP MEANS MOVING OUT OF HOME
Sponge crabs carefully hollow out pieces of sponge to snugly fit their dome-shaped bodies, but every time they moult they grow larger and the work of making a new home has to start again.

HARLEQUIN SWIMMING CRAB
Lissocarcinus laevis

FAMILY:	Portunidae
OTHER NAMES:	Anemone crab
DISTRIBUTION:	Widespread Indo-West Pacific; from South Africa to Japan, Australia, Hawaii and French Polynesia
HABITAT:	Inshore reefs in association with sea anemones; sublittoral to 85 m (280 ft) depth
FEEDING HABITS:	Predator
NOTES:	Easily recognizable when alive by its characteristic colour pattern, although colours can vary in hue
SIZE:	To about 20 mm (¾ in) carapace width

CRABS OF THE 'SWIMMING CRAB' FAMILY Portunidae are immediately recognizable by the flattened paddles on the ends of the last pair of legs. These 'natatorial legs' make them highly efficient swimmers, and in combination with strong and sharply toothed claws, make them fast and deadly predators. Species of *Lissocarcinus* are somewhat unusual amongst portunids because most seem to prefer to live in a symbiotic relationship with other invertebrates, in particular a variety of sea cucumbers, sand dollars, some corals, salps and sea anemones.

SHARING A LIFE

Sometimes species unite for mutual benefit, one providing shelter and safety while the other defends its host from predators or does the housekeeping by removing detritus and potential pathogens from the host's surface. In other cases, there may be no obvious benefit to either, just a marriage of convenience! The nature of the relationships that *Lissocarcinus* species have with their hosts has not been well studied. One suspects that the crabs get most of the benefit. The harlequin crab mostly lives on sea anemones, but it can sometimes turn up randomly on the reef as well, so the relationship does not seem to be obligatory, although perhaps they simply venture away on feeding forays or in search of mates, and return to their anemones later.

TAKING SHELTER
Harlequin swimming crabs are perfectly at home amongst the venomous tentacles of sea anemones. This one was photographed on Sabang Beach, Mindoro, in the Philippines.

SCULPTURED CRAB
Vultocinus anfractus

THIS CRAB CREATED MUCH EXCITEMENT when first discovered in the Philippines in 2007. Not only was it a new species in a new genus, but it could not be placed in any existing family either. Molecular DNA studies have since shown that it appears to be an ancient lineage not closely related to other Goneplacoidea species to which it is otherwise most similar. A 2020 research paper added to the story with the discovery of some well-preserved fossils – *Pyreneplax basaensis* – from the upper Eocene (around 56 mya) of northern Spain. This extinct species is also clearly in the family Vultocinidae, and its similarity to numerous other Eocene genera suggests that vultocinids may once have been quite diverse. In that case, *Vultocinus anfractus* could be the last survivor of a group once common in ancient seas.

ECOLOGICAL ENIGMA

While this is a fascinating crab from an evolutionary perspective, it is known from very few specimens and its ecology is poorly understood. Two of the original crabs were living inside cavities in sunken wood, their colour matching the substrate perfectly. Their spinous lower leg margins helped to anchor them against crevices, and their dense coating of hairs was also thickly coated in detritus and mud, adding to their camouflage. However, later specimens from New Caledonia and Vanuatu were obtained from general reef collections, so the crabs in the Philippines may have just been making good use of an abundant local resource.

FAMILY:	Vultocinidae
DISTRIBUTION:	Central West Pacific; only known from the Philippines, Vanuatu and New Caledonia
HABITAT:	30–200 m (100–650 ft)
FEEDING HABITS:	Unknown, but presumed to be a predator of other small invertebrates, based on its claw shape and armature
NOTES:	The single new species and new family were first described as recently as 2007
SIZE:	To about 30 mm (1 1/8 in) carapace width

THE IMPORTANCE OF CONCEALMENT

One of the first specimens was found living inside a piece of wood at 200 m (650 ft) depth – a camouflage so perfect that it seemed like a specialized habitat. Subsequent collections, however, suggest that it probably lives hidden in a variety of crevices or cavities in suitable reef habitats.

ADAMS ZEBRA CRAB
Zebrida adamsii

FAMILY:	Pilumnidae
OTHER NAMES:	Urchin zebra crab
DISTRIBUTION:	Central Indo-West Pacific: eastern India to Japan and south to northern Australia
HABITAT:	Commensal on sea urchins
FEEDING HABITS:	Parasites or micro-predators that eat the urchin's tube feet and epithelial skin tissues
NOTES:	Often shares the host with a pair of similarly coloured small shrimps (*Periclimenes colemani*)
SIZE:	To about 20 mm (3/4 in) carapace length

ZEBRIDA ADAMSII BELONGS TO A SPECIALIZED subfamily of the pilumnid reef crabs, Eumedoninae – all obligate symbionts of Indo-West Pacific echinoderms. Most species live on the surface of their hosts, although two have taken up residence inside. Females of *Echinoecus pentagonus* live in calcified galls in the rectum of urchins, and *Hapalonotus pinnotheroides* lives in a fleshy cyst inside the cloaca of sea cucumbers. *Zebrida adamsii*, with its long spines and distinctive vertical stripes, is one of the most instantly recognizable and photographed.

NO HARM MEANT

Zebrida is relatively indiscriminate regarding the urchins it chooses to live on, having been found on at least ten species that vary greatly in shape and spination, including the highly venomous flower-urchins (*Toxopneustes* species). The crabs live between the spines of the urchin, gripping them using the hook-shaped ends of their legs. While many such relationships are mutually beneficial, in this case the crabs are parasites. They feed on the urchin's tube feet and epithelial skin tissues, although they rarely do much long-term harm. Females like to stay safe at home on their chosen urchin, while the slightly smaller males roam the reef looking for females to visit for mating. This system has been referred to as 'pure-search polygynandry of sedentary females', or more simply, 'visiting girlfriends'!

PEACEFUL COEXISTENCE

Zebrida adamsii is named for its characteristic brown and white zebra stripes. They live between the protective spines of a variety of echinoids (sea urchins), upon which they also feed. It seems the urchins gain no advantage from their presence, but the crabs rarely do much harm either.

4

REPRODUCTION, COGNITION AND BEHAVIOUR

REPRODUCTION

WHILE REPRODUCTION IS ESSENTIALLY a physiological process (mentioned in Chapter 2), crabs have behavioural strategies to ensure maximum survival of the new generation. Many male crabs have evolved elaborate courtship rituals, and females in turn have developed extraordinary behaviours to ensure their eggs and larvae get the best start in life, even to the extent of advanced levels of maternal care.

COURTSHIP AND MATING

For most marine crabs, courtship is a perfunctory affair. Actual mating involves the male rolling the female onto her back, both crabs flex their pleons away from their bodies, and he inserts his twin first gonopods into her corresponding vulvae for internal fertilization (in Eubrachyura), or into a seminal receptacle for later external fertilization (in Podotremata and Archaeobrachyura; see page 72). Mating in many cases takes place when the female crab is still soft from moulting. However, despite it appearing as if she is being subservient to the males advances, the female may well be controlling the whole affair. The helmet crab (*Telmessus cheiragonus*; family Cheiragonidae) has mating linked directly to moulting. Females produce two different pheromones that strongly affect their new partner. The first induces the male to begin pre-moult guarding (protecting her with his body and claws), while the second induces copulation and post-moult guarding until she is able to harden her shell.

In contrast, many of the more socially advanced intertidal crabs, particularly in the family Ocypodidae, have taken courtship behaviour to a much more sophisticated level. There has been a great deal of study on the mating rituals of fiddler crabs (*Uca* and related genera). Fiddler crabs take sexual dimorphism (differences between the sexes) to an extreme. While the female crabs have two dainty and diminutive claws that they use during low tide to sift the mud surface for food, the male has only one such tiny claw – the other is an ungainly, brightly coloured monster that can weigh half as much as the crab itself. The male uses this giant claw to attract the attention of females by both drumming on the mud and performing intricate waving displays. He also uses it to defend the territory around his burrow, aggressively pushing away male interlopers (although it is typically only pride that gets hurt).

LEFT: Many male crabs stand guard over their chosen female until she moults, her shell is soft, and she is ready to mate. This ridged swimming crab (*Charybdis natator*) is carefully rolling the female onto her back into mating position.

OPPOSITE: A male fiddler crab (*Tubuca polita*) in mid-display. Fiddlers will display to females for courtship, or as acts of aggression against intruding males that may want to steal their burrows.

STALK-EYED CRAB WAVING DISPLAY

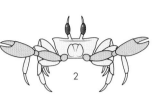
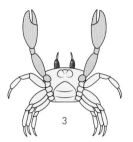

Each species of fiddler crab has its own special display to which only females of their own species will respond – and it seems that females can be very discriminating indeed. Firstly, size does matter! The males with the largest claws seem to have the most success. Secondly, females seem much more attracted to fast wavers, perhaps because speed is a good indicator of strength and vitality, and therefore breeding stock. If the female is interested in mating, she will approach the male and stroke him using the legs on one side of her body. In many fiddler species, females will then move to the burrows of their chosen males for both mating and subsequent egg incubation; however, in many others, mating will take place on the surface before females return to their own familiar burrow surroundings.

Recent research has found that the big, heavy claw of male fiddler crabs requires a lot of energy to move, especially while the crab is exposed on a hot tropical mud bank. However,

ABOVE: Male stalk-eyed crabs (*Tmethypocoelis ceratophora*, see page 172) have a unique waving display. They begin with the claws held in front of the face (1), then unfold them sideways (2). Next, they lift them up to their full extent, raising their body and lifting the first pair of walking legs off the ground (3). Both claws are then pointed out in front of the crab before being refolded (4), and beginning again (5).

the physiological stress incurred is offset, to a great degree, by heat loss – the claw functioning somewhat like a radiator, with heat from the body being dissipated into the air via convective heat transfer. The crabs can then remain on the surface for longer, giving them more time for their active social engagements, and for foraging – especially important considering they have only one small claw for feeding.

REPRODUCTION 145

MIGRATIONS

Some crabs congregate in huge numbers at certain times of the year. For example, adult female 'land crabs' of the family Gecarcinidae must migrate to the sea in time for their eggs to hatch and the larvae to be released. These migrations can involve huge numbers, with crabs entering houses and disrupting traffic on coastal roads. *Gecarcoidea natalis* on Christmas Island, *Gecarcinus ruricola* in Cuba and the San Andres Archipelago off Columbia, and *Johngarthia lagostoma* on Ascension Island all undertake spectacular migrations. Migrations usually begin with the onset of the summer rains, and therefore vary in timing from year to year. The Christmas Island red crab breeding and spawning event is the most famous, involving millions of crabs (see pages 200–1).

The commercially fished tanner crab (*Chionoecetes bairdi*) also forms huge springtime breeding aggregations, but in deep water (150–200 m [500–650 ft]) off Alaska, in the US. Over 200 pods with more than 100,000 individuals have been reported over an area of just 2 hectares (5 acres). The pods are made up entirely of females carrying eggs with late-stage embryos, and they release their larvae in synchrony, on the full moon marking the highest spring tide. Afterwards the pods quickly disperse.

Spider crabs are famous for annual migrations to form massive aggregations. Depending on the species, this phenomenon also seems to have other motives, including communal moulting, and mating opportunities. In Port Phillip Bay near Melbourne, Australia, great spider crabs (*Leptomithrax gaimardii*) can gather by their tens of thousands, forming a moving wave of bodies that can be 2 m (6½ ft) high and more than a kilometre long. Similarly, the European spider crab (*Maja brachydactyla*) undertakes a summer migration into the shallow waters off the coast of Galicia, Spain, where it forms mounds or pods ranging in size from a few dozen individuals, up to 50,000. In both cases, the crabs are protecting each other during a critical stage of their life cycle; studies on *Maja* have shown that the crabs on the surface of the pods still have hard shells, whereas those in the centre are just beginning or finishing their moulting, and so are soft and vulnerable to a hungry predator.

OPPOSITE: Each year Gaimard's spider crabs (*Leptomithrax gaimardii*) migrate into shallow waters off southern Australia to form immense winter aggregations (see page 126).

RIGHT: Christmas Island red crabs are famous for their annual shore breeding migration to release their eggs. Long, sharply pointed legs allow them to climb over virtually anything in their way.

RAISING THE FAMILY

Female brachyurans carry their developing eggs attached to their pleopods under the pleon, where the embryos are brooded until hatching. The female is much more than a passive egg carrier, however. Looking after the eggs is a full-time role, and egg-bearing females show advanced levels of maternal care. The developing embryos are encased in a semi-permeable egg membrane that allows them to absorb oxygen from the water, so the mother crab must make sure her egg broods are ventilated with plenty of fresh water by undertaking 'pleonal flapping' to keep water flowing over them. As the embryos grow, the nascent crabs need more and more oxygen, so the female must gradually increase her flapping rates. The eggs are also very delicate, and must be kept clean and free of infections by bacteria, filamentous fungi and protozoans, so females regularly groom their eggs using their chelae and other legs.

BELOW: A harlequin crab (*Lissocarcinus laevis*) fiercely guards its large clutch of orange eggs. Gravid (pregnant) female crabs are often said to be 'berried'.

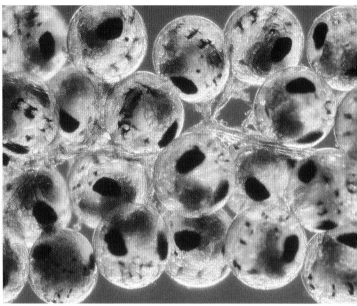

ABOVE LEFT: *Calvactaea tumida* lives commensally with carnation corals. The huge number of tiny eggs will give its larvae a better chance to find a suitable host.

ABOVE: Eggs (here, from the brown land crab) develop under the pleon until ready to hatch into the first zoeal stage and swim free in the plankton.

Terrestrial and intertidal crab mothers also face other physical challenges that can be deadly to their sensitive developing eggs (such as overheating and desiccation), and thus motherhood can have major impacts on their own lifestyles. For example, most egg-bearing females of the beach ghost crabs (*Ocypode* species) and the muddy shore fiddler crabs (*Uca* species) stay in their burrows for extended periods because this gives them access to water, and helps them avoid exposure to both solar ultraviolet radiation, and the high temperatures on the surface. The terrestrial Christmas Island red crab (*Gecarcoidea natalis*) uses a different strategy. Mating for these crabs only occurs in the specially prepared 'nursery' burrows created by the males after they have migrated to the edge of the coast. Once mated, females stay in the safety and cool of the burrows until the eggs are ready to hatch. Very importantly, the eggs and embryos have also evolved to be able to absorb atmospheric oxygen directly.

One genus of false spider crab, *Neorhynchoplax* (family Hymenosomatidae), is unique amongst brachyurans for developing ovovivipary – eggs that are hatched within the body of the parent. The internal skeleton of the body is compressed so as to leave a large empty space in which the fertilized eggs develop. The few, large eggs are ventilated via an internal connection with the branchial chamber, and are finally released to hatch not through the vulvae as in other crabs, but via a tear on the sternal membrane – something like a self-induced caesarean!

The timing of egg release can be very important for survival of the young. Many coastal crabs follow lunar cycles, with larval release synchronized with the new or full moon; the strong spring tides that occur at this time quickly pull larvae away from the shore. In contrast, some crabs (like *Carcinus maenas*) release on neap tides, when the tidal currents are at their weakest, keeping the larvae close to a favourable habitat where their chances of survival will be the greatest. Regardless of tide, the simultaneous release of vast numbers of larvae from many crabs serves to satiate the appetites of any predators waiting in shallow water, and increase the odds that the remaining larvae will survive.

LIFE CYCLE OF A CRAB

A typical life cycle for a crab, as exemplified here by the giant mud crab (*Scylla serrata*), includes five or more planktonic zoeal stages and a megalopa that settles to the bottom and moults into the first crab stage.

PLANKTONIC LARVAL DEVELOPMENT

The number of eggs that a crab produces varies dramatically. There can be fewer than 20 in tiny false spider crabs, or as many as 6 million in some large portunid swimming crabs such as *Scylla serrata*. It all depends on the breeding 'strategy': produce large numbers of small zoea that can be spread far and wide, and the strongest and luckiest survive (called r-selection); or invest a lot of energy in producing a small number of offspring in the hope that most will survive (K-selection). K-selection is more common in stable environments where it is best to prevent juveniles straying too far from home, whereas r-selection bestows the ability to quickly spread into new areas.

Once embryos are fully developed, most crabs release them into the sea, where they will begin the zoeal (or larval) stage of their life. As members of the plankton community, larval crabs feed and continue to grow, moulting through a number of similar-looking zoeal stages, with each stage developing more complex limbs for feeding and swimming. The majority of brachyuran families have four or five zoeal stages, but this can vary; the majoid spider and decorator crabs have only two stages, the false spider crabs (family Hymenosomatidae) have three, most of the swimming crabs (family Portunidae) have five or six, and intertidal grapsid crabs can have up to eight! Overall, the planktonic development stage can last from two to three weeks to a couple of months. The final zoeal moult sees the first major metamorphosis into a bottom-dwelling 'megalopa' stage that has fully functional pereiopods and pleopods. The megalopa then moults into the 'first-stage crab', which is just an immature-looking juvenile of the adult it is destined to become.

RIGHT: A megalopa – the bottom settlement stage marking the end of planktonic larval development. At the next moult, the pleon will tuck under the body and the adult crab shape will be obvious.

Crab larvae can play an active role as to where they settle. Coral reefs are noisy with life, and experiments have shown that larvae of crabs that live on the reef swam to traps with reef sounds, while non-reefal species avoided them in preference for quiet traps. Similarly, metamorphosis to a megalopa can be accelerated by the smell of adults of their own species, and chemical cues from symbiotic hosts or bacterial biofilms that grow on the adult habitat.

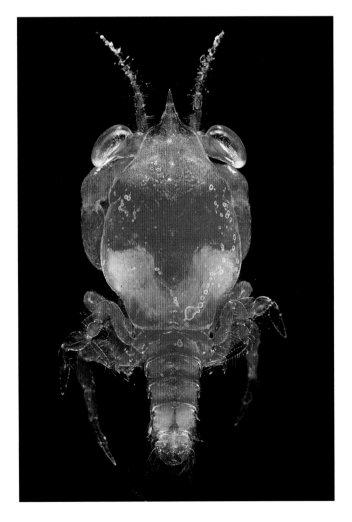

BELOW: Juvenile crabs may also undergo mass migrations back to the habitat that will nurture them into adulthood. The return of baby Christmas Island red crabs about a month after eggs are released is almost as amazing as the seaward migrations of their parents.

When adults live in fresh water or on land, either the megalopae or the juvenile crabs must migrate upstream or across country. Perhaps the most remarkable migrations occur in the mitten crabs (*Eriocheir* species), where the megalopae migrate upstream through the estuaries, moult into juvenile crabs, and travel up to 1,500 km (900 miles) over a period of several years. Somewhat more visible juvenile migrations occur in the gecarcinid land crabs, with the baby Christmas Island red crabs (*Gecarcoidea natalis*) forming a red carpet over the rocks and beaches as they leave the sea for the relative safety of the forests beyond.

ABBREVIATED DEVELOPMENT

While free-living zoea are the basis for the majority of crab life cycles, many crabs, especially those living in fresh water, have eliminated most of the zoeal stages; instead, development occurs mostly or fully within the egg, much like reptiles and birds. These species typically have smaller broods of large eggs that are rich in yolk stores that the young crab needs throughout its development (called 'lecithotrophy'). This, of course, results in the mother birthing numerous baby crabs, and some serious maternal-care strategies have evolved to protect and help the youngsters survive.

Such abbreviated development has independently evolved numerous times in different crab groups, even in some marine species, but it is most common in freshwater or terrestrial groups. Curiously, it has even evolved in the typically marine family Chasmocarcinidae, which has one freshwater subfamily, Trogloplacinae, with only two genera (*Australocarcinus* and *Troglocarcinus*) and five species. These crabs each live in isolated streams and caves in the tropical central West Pacific and the Seychelles, and have large eggs that hatch directly into megalopae. These remain with their mother until they develop at least into the first crab stage and are large enough to fend for themselves.

SPECIAL MATERNAL CARE

Most families of 'true' freshwater crabs typically all have direct development. The egg-incubation period is often around 30 to 45 days, and the juvenile crabs remain attached to the female from two weeks up to four months after hatching, depending on the arrival of monsoonal rains. The desert-adapted Australian inland crab (*Austrothelphusa transversa*) remains dormant ('estivates') in underground burrows while waiting for summer rain. An estivating female crab containing 42 first-stage crablets under the pleon has recently been reported, indicating that egg development progresses underground during the dry season while the mother sleeps. The young crabs are then able to hit the ground running when wet weather arrives and the streams start to flow.

The mandarin crab (*Geosesarma notophorum*) is a terrestrial crab from Sumatra that also has direct development, and a different twist on maternal nurturing. It broods just a few large eggs, and after hatching, the baby crabs roam over the back of the female until ready to become independent.

The bromeliad crab (*Metopaulias depressus*; family Sesarmidae) occurs in western and central Jamaica, and raises its young in the water-filled leaf axils of bromeliads. The females of this species are arguably the best mothers a crab could ever hope for. The female chooses a plant that holds large volumes of water, then carefully prepares it as a nursery. She removes any dead leaves or organic detritus that will use up oxygen, circulates the water to oxygenate it, then places empty snail shells into the water to provide a calcium

BELOW LEFT: An adult female mandarin crab in the Lingga Islands, Sumatra, Indonesia. This crab broods just 8 to 12 large eggs at a time. The baby crabs perch on top of her carapace.

BELOW: The bromeliad crab lives and breeds on bromeliads that grow in Jamaica. It shows advanced maternal care behaviour akin to that of birds, and would put many a vertebrate to shame.

A CUCKOO IN THE NEST: THE STRANGE CASE OF SACCULINA

Sometimes a female crab with a mass of eggs is not what it seems. Both male and female crabs are prone to being parasitized by very aberrant forms of barnacle known as rhizocephalans. The barnacle grows inside the host's body, hidden until it is ready to produce its own eggs. At this point it produces a large, leathery sac ('externa') that protrudes from beneath the crab's tail flap, which is where its eggs develop. The particularly macabre aspect of this relationship is that the barnacle secretes hormones which, in the case of a male host, feminize the crab so that it develops into what looks like a female, capable of nurturing the egg mass as its own. With a female host, they lead to castration of the crab's own reproductive organs, so after caring for the barnacle's eggs, she dies before ever becoming a mother herself. Infection rates are typically low (1–2 per cent), but in some areas can be a concern for fisheries management. There are numerous types of rhizocephalans, and not all produce eggs under the tail flap; in some cases, clusters of small grape-like egg bodies protrude from the joints of legs instead.

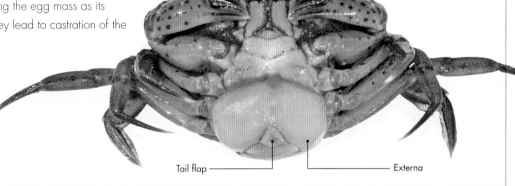

RIGHT: A green shore crab (*Carcinus maenas*) that has been infected with an internal parasitic barnacle, *Sacculina carcini*.

Tail flap — Externa

source for strong young shells, and act as a pH buffer. She produces up to 90 babies, with the second zoeal stage moulting directly to a juvenile crab. But her work does not stop there. The babies need to be fed, so like a mother bird, she keeps busy capturing insects such as cockroaches and millipedes to feed her ravenous offspring. She also keeps constant guard against predators looking for an easy meal, particularly spiders and aggressive damselfly nymphs that have also evolved to take advantage of bromeliad pools – just one nymph can eat up to five crablets a day. This nursery can be maintained for several months, and even be home to successive broods of eggs, so that a multigenerational family group is often living together. With all this care lavished on them, it is not much wonder that the kids are reluctant to leave home, but they do eventually vacate when they are large enough to properly defend themselves.

The related Jamaican snail-shell crab (*Sesarma jarvisi*) has a similar story of devoted motherly care. Instead of bromeliads, this crab breeds in large, empty land-snail shells; the crab turns the shells upside down to collect rainwater, but will also carry water to the shell to keep it topped up.

BEHAVIOUR AND INTELLIGENCE

WHILE IT MAY NOT APPEAR SO to the casual observer, crabs navigate their lives with purpose – even if that purpose is driven by primal needs. Most crabs live in complex environments with many hidden dangers, be it under the sea or in a tropical forest. Therefore it is vitally important to learn the location of, and routes to, critical resources. It is also a good idea to remember which hole the moray eel lives in, and under which rocks the octopus hides.

The mangrove tree crab (*Parasesarma leptosoma*) traverses an intricate three-dimensional habitat amongst the roots and canopy of East African mangrove swamps. Experiments using an artificial model of one of its major predator crab species, a forceps crab (*Epixanthus dentatus*), proved that the *Parasesarma* could recognize the shape of its enemy and thus avoid the potential danger. Similarly, the predatory Indo-West Pacific swimming crab *Thranita crenata* is able to navigate across sandy-mud flats using complex landmark memories, an ability that at least rivals the extraordinary navigational sophistication of honeybees.

Such intelligence has also been demonstrated in the European shore crab (*Carcinus maenas*), an important predator and scavenger in intertidal and shallow subtidal environments. It became the subject for an experiment investigating how well crabs could learn. The testing ground was a complex multi-turn maze (similar to the type used to test mice) with some fresh, juicy mussel meat at the end. On their initial attempt the crabs found the treat only after lots of trial and error, often taking over an hour. However, after only four tries at the labyrinth, once a week, most crabs were going straight to the end in under eight minutes, without any wrong turns. Not only that, but after an absence of two weeks the crabs still had the route memorized and would skip directly to the end of the maze, even when the reward was no longer there to give them sensory incentive. So do not underestimate a crustacean's brain.

ABOVE: Good grooming is essential. An American Atlantic sand fiddler (*Leptuca pugilator*) uses the brush of setae on the palp of its third maxilliped to clean its eyes.

CRAB TALK

One does not normally think of crabs as being chatty, but research is finding that many crabs have developed a significant vocabulary, and have a lot to say! It has been known for a long time that crabs are able to stridulate in a way similar to cicadas. Stridulation is the simple process of rubbing together two body parts. The fixed 'rasp' is typically composed of one or more rows of ridges, granules, beads or short, rigid setae. The 'plectrum', which is typically a single raised ridge or sharp edge, is rubbed across the rasp to produce a sound. The size and spacing of the rasp's ridging, combined with the speed of the plectrum, produces different tones and sound frequencies.

Male fiddler crabs and ghost crabs (family Ocypodidae) appear to be unique amongst Crustacea because, during the breeding season, they actively call out to attract females to come to their special 'mating' burrows. The ghost crabs

(*Ocypode* species) are well known to stridulate with distinctive rasps on the inside of their claws, and in fact some large crabs can be so loud that their sound can carry up to 10 m (33 ft) away. In contrast, fiddler crabs (*Uca* species) do not stridulate, but call by 'rapping' (rapidly vibrating their large claw against the ground) or 'honking' (making low-frequency sounds by moving their walking legs up and down rapidly). Some species may also make one to four brief, higher-frequency pulses by rapidly tapping individual legs against the substrate. Sixteen different movement patterns for sound production have been documented in *Uca*.

One species of ghost crab, *Ocypode quadrata*, has an even more remarkable ability: it can make a loud internal rasping by using the gastic mill in its stomach. A pair of ridged plates rub against a long central tooth to produce stridulation. This amazing form of gastric vocalization also occurs in the New Zealand paddle crab (*Ovalipes catharus*). New findings indicate that this species has very sophisticated communication, and can make at least three distinct sounds: the 'rasp', 'zip' and 'bass'. The rasp is made by conventional stridulation (rubbing a leg against the claw) and is produced intermittently by both sexes, but dramatically increases in the presence of food. Playing the recorded sound to other crabs immediately elicits food-foraging behaviour, suggesting that when they hear their neighbours have found food, they home in on it themselves. The bigger the crab, the lower the sound frequency and pitch, so crabs can judge each other's size by their 'voice', and so get a sense of the competition they will face. The zip and bass sounds originate (as in *Ocypode*) from tooth grinding in the gastric mill, but are only made by adult males during courtship behaviour with a receptive female, and during post-copulatory mate-guarding. It appears to be a message to other males to 'stay away from my girl'.

While crabs seem to be good at making a variety of sounds, the mechanisms they use for hearing are still poorly understood. A variety of external and internal sensory receptors are probably all involved. One organ in particular – Barth's organ – is possessed by many crabs, and appears particularly sensitive to vibration. It is found at the base of the merus on each walking leg, and takes the form of a small, thin-walled, depressed 'window'. The ligaments and muscle attached to the inside of this window are embedded with neurons that respond to mechanical vibrations of the window – a system not unlike the human eardrum.

BELOW: Some crabs make sounds to communicate by using a variety of specialized stridulatory structures unique to each species. Stridulation is the act of rubbing two body parts together – one has a rasp (one or more rows of ridges, granules or beads); the other a raised plectrum. Mangrove shore crabs of the genus *Parasesarma* turn one claw to point downward, while quickly rubbing up and down against it the large granules on top of the finger of its opposing claw.

STRIDULATION USING THE CLAWS

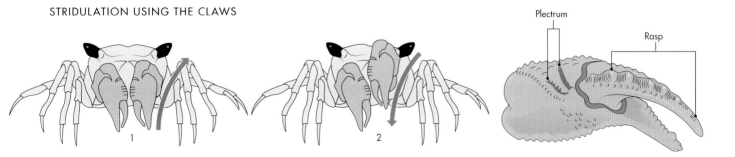

RIGHT: Two male painted ghost crabs (*Ocypode gaudichaudii*) fighting. This species occurs on beaches in the East Pacific, from El Salvador to Chile and the Galápagos Islands.

CRABS WITH ATTITUDE

Many crabs, especially those based in intertidal burrows, live in close communities, and have developed a surprisingly complex social system with dominance hierarchies. This often involves male competition for the attention of females using sophisticated mating dances, but it seems that the wrestling and clash of claws between two male crabs is not simply about pushing and shoving. In certain crabs at least, the winner positively gloats in a way that is eerily similar to human fist-pumping! After two male Southeast Asian mangrove shore crabs (*Parasesarma eumolpe*) tussle over a female, the winner will often perform a taunting victory dance to intimidate and belittle the loser into leaving the fray (which it mostly does). This is performed by turning one claw to point downward, while quickly rubbing the other claw up and down against it – probably also making a defiant stridulating call using the row of large granules on the top of its finger. And as mentioned, other crabs can also produce a 'stomach growl' as a warning to stay away.

HIDE AND SEEK

Hiding from predators is an important aspect of crab behaviour. However, hiding all day is not an option when there is food to be gathered, so it is not surprising that camouflage and other forms of active concealment have evolved. While more typically a defence strategy, some predators, especially those that specialize in ambush attacks, are also expert at camouflage. The whole purpose of visual deception is to be nearly undetectable, and thus ignored by predator or prey alike.

ABOVE: The carapace of the reef rubble crab (*Aethra scruposa*) looks just like a piece of coral rock, particularly once barnacles, bryozoans and tube worms begin to grow on top of its shell.

BELOW: A conical gorgonia crab (*Xenocarcinus conicus*) is perfectly camouflaged on this red gorgoniid fan coral – tiny splashes of yellow on its body and legs even match the colour of the coral polyps.

Where crabs are unique from almost all other animals is in their specialized behaviours of carrying and self-decoration. In a very real sense, carrying and decorating behaviour must be considered tool use. The only similar behaviour amongst invertebrates is the jealous possession and use of coconut shell-halves as shelters by the veined octopus of Indonesia.

CRYPSIS AND MIMICRY

Many crabs employ 'crypsis', meaning that they become hidden against their background. Species of *Clistocoeloma* (Varunidae) are intertidal mangrove crabs that are not only mud-coloured, but have a fur of short hairs (setae) that acts as a magnet for the passive accumulation of silt, further obscuring their presence. Other species have deeply eroded and pitted shells that look like the rubble bottoms on which they live, or are shaped like dead fragments of calcified green alga, or foraminiferan discs. The plate crab (*Aethra scruposa*) has taken on the appearance of a dead coral plate, with its legs and claws hidden beneath its broad round shell.

Mimicry (or masquerade), is another aspect of crypsis more typically used for when a crab pretends to be something else through shape, colour and behaviour. This is particularly common amongst crabs that live as external symbionts on coral or other invertebrates, often achieving remarkably bright colours and patterning to match that of their host, rendering them almost invisible. The spider crabs (superfamily Majoidea) include some great mimics (see page 208). Similarly, species of *Xenocarcinus* take the colour of their fan- and black-coral hosts, even to the extent of mimicking the different colours and shapes of the coral polyps. The carapace and pretty green colour of the arrowhead crab (*Huenia heraldica*; family Epialtidae) perfectly mimics the living green *Halimeda* alga around which it lives, but as one of the decorator crabs it also attaches pieces of the same alga to its carapace to create the perfect disguise.

DRESSING UP: DECORATING BEHAVIOUR

Certain spider crab families (majoids) are famous for their varied and highly individualized habit of decorating their upper body and legs in an effort to conceal their outlines. Popularly called spider crabs, decorator crabs or masking crabs, these are the only known brachyurans to show such behaviour. Generally, crabs spend a lot of time grooming themselves to stay clean, but many majoids use their very flexible claws to cover themselves in a variety of living marine organisms (see page 168). Thus camouflaged against their background, they are able to graze safely for food across the bottom, unnoticed by predators. Given that every crab looks different, it also becomes very difficult for a predator to ever develop a consistent prey search pattern.

LEFT: This decorator crab (a *Hyastenus* species) has covered itself in a living coat of hydrozoan polps.

OPPOSITE LEFT: Almost invisible against a background of encrusting lace coral, *Crossotonotus spinipes* belongs to the small family Crossotonotidae, which are typically coral reef inhabitants.

OPPOSITE RIGHT: An upturned sponge crab, *Mclaydromia dubia*, showing the snug-fitting home that it has hollowed out underneath a piece of sponge.

CARRIERS

Carrying behaviour is unique to the oldest and most primitive of crabs, the podotremes and archaeobrachyurans, and was first used in ancient seas over 180 million years ago. It requires special modification of the last pair, or last two pairs, of legs to be able to grip and carry an object. These 'carrier' legs are no longer useful for walking, but are agile and able to fold back over the back of the carapace, and the last two segments are modified into specialized claws.

Carrier crabs select suitable objects from the environment and use them to cover and protect themselves from predators. These objects are typically anything that they can find, and can thus be highly varied, including dead shells, living sponges or small colonies of compound ascidians. Sometimes this mobile camouflage is not just passive cover – stinging sea anemones or spiny sea urchins can become formidable defensive allies.

Sponge crabs (family Dromiidae) have perhaps the most highly evolved carrying behaviour. They hollow out pieces of sponge to fit snugly over their whole upper body (see also pages 134–35) and become impossible to see, with only movement giving away their position.

In contrast to sponge crabs, which only change their sponges when they moult, carrier or porter crabs (Dorippidae) seem happy to swap their cover as necessary. The Australian leaf-porter crab (*Paradorippe australiensis*) lives in muddy estuaries and carries a mangrove leaf on its back – a hungry fish is not likely to be interested in a leaf seemingly 'floating' slowly across the bottom. Leaves may not last long, but they are in plentiful supply. The granulated porter crab (*Paradorippe granulata*), a species from China and Japan, prefers to carry shells that have one, two or even three sea anemones attached, for that extra bit of protection.

Some porter crabs have a dramatic association with sea urchins, starfish, anemones, and even jellyfish. For example, the rough porter crab (*Dorippe quadridens*) often carries urchins many time its size, even holding on to them as it buries itself just below the sand while it remains inactive during the day – presumably the urchin would much prefer its freedom! By contrast, the mauve porter crab (*Dorippoides facchino*) seems to have a special arrangement with a sea anemone; the anemone secretes a thin plate under its trunk, offering the crab something to grip, and the two form an intimate relationship, gradually growing in size together.

STRIPED BOX CRAB
Calappa lophos

FAMILY:	Calappidae
OTHER NAMES:	Common box crab, common shame-faced crab
DISTRIBUTION:	Eastern Indian Ocean to western Pacific; north to Japan and south to Australia
HABITAT:	Sand, mud and gravel bottoms; 5–140 m (16–460 ft) depth
FEEDING HABITS:	Predator: molluscs, hermit crabs and other invertebrates
NOTES:	Sometimes sold for food in markets in the Philippines and East Asia; mainly caught by trawlers, tangle nets and sometimes traps
SIZE:	To about 10 cm (3 7/8 in) carapace width

THE STRIPED BOX CRAB is one of 42 species in the genus *Calappa*. Collectively known as box crabs, they look like large rocks or lumps of coral, especially when at rest or partially buried in sand or rubble. In Asian countries they also are referred to as 'shame-faced crabs' because they completely hide their 'faces' behind large, deep and flattened claws that fit perfectly against the edge of the carapace; similarly, their short, slender walking legs are usually totally hidden under wing-like expansions from the back of the shell. Despite being superbly camouflaged, they are one of the most distinctive crabs found in tropical and subtropical seas. *Calappa lophos* is one of the most common, but also one of the prettiest species, with a pinkish to yellowish-cream shell covered in small red flecks, and distinctive thick blood-red spots and lines on the chelipeds and sides of the carapace.

MARINE 'TIN OPENERS'

Calappa species are specialist feeders on molluscs and have developed a special 'shell-opener' on their right claw, used to break snails out of their hard, coiled casing. The shell is held with the pointed end away from their body, and the edge of the aperture is hooked under an extra 'tooth' on the outside of its right claw. The crab then systematically breaks away the shell as it rotates it, until it reaches the snail hiding inside. The victim is then pulled out with the sharply pointed fingers of its smaller left claw (see page 66).

SPAWNING

The water turns milky around this large female box crab as her eggs hatch and her convulsing pleon releases a million tiny larval zoea into the water to live and grow in the surface plankton. A few lucky ones will avoid being eaten by the teaming mouths of larval fish and other larger predators, and eventually settle again to the bottom to begin the cycle once more.

CHRISTMAS ISLAND RED CRAB
Gecarcoidea natalis

THE ANNUAL CHRISTMAS ISLAND red crab breeding and spawning event is one of the most spectacular animal migrations on Earth, involving at least 50 million crabs. It begins with the onset of the summer monsoonal rains around year's end. Males migrate first, to excavate special 'mating burrows' near the shore. Females stay in these burrows for about two weeks, nursing around 100,000 fertilized eggs under their flap-like abdomens, before emerging en masse to spawn into the sea, always just before dawn during the last quarter of the moon, just as high tide begins to recede.

A PERILOUS JOURNEY

The hatched larvae swim, feed and grow in the rich tropical plankton before moulting into tiny bottom-dwelling megalopae. These emerge from the sea, and in turn moult into minute baby crabs that swarm over the rocks on the arduous trek back to their forest home (see pages 152–53). Populations have almost halved over recent decades from around 120 million to 50 or 60 million, largely due to the attack of huge infestations of the invasive yellow crazy ant. Management measures are under way to control these terrible pests.

FAMILY:	Gecarcinidae
OTHER NAMES:	Red crab
DISTRIBUTION:	Indigenous to Christmas Island, northeast Indian Ocean
HABITAT:	Most common in the moist island rainforests, but can be found everywhere, even in town gardens
FEEDING HABITS:	Primarily herbivorous, eating fresh or fallen leaves, fruits, flowers and seedlings; will eat other dead crabs or birds, given the opportunity
NOTES:	By sheer biomass, this one species surpasses that of all animal life in many of the world's rainforests! Red crabs will drown if forced to stay under water.
SIZE:	Up to 12 cm (4¾ in) carapace width

HOME FROM THE SEA
Back from their gestation in the sea, these tiny newly hatched crablets swarm over the shore, seeking the shelter of the moist, shaded inland forests. Predators abound on the exposed limestone terraces that surround the island. Many will perish.

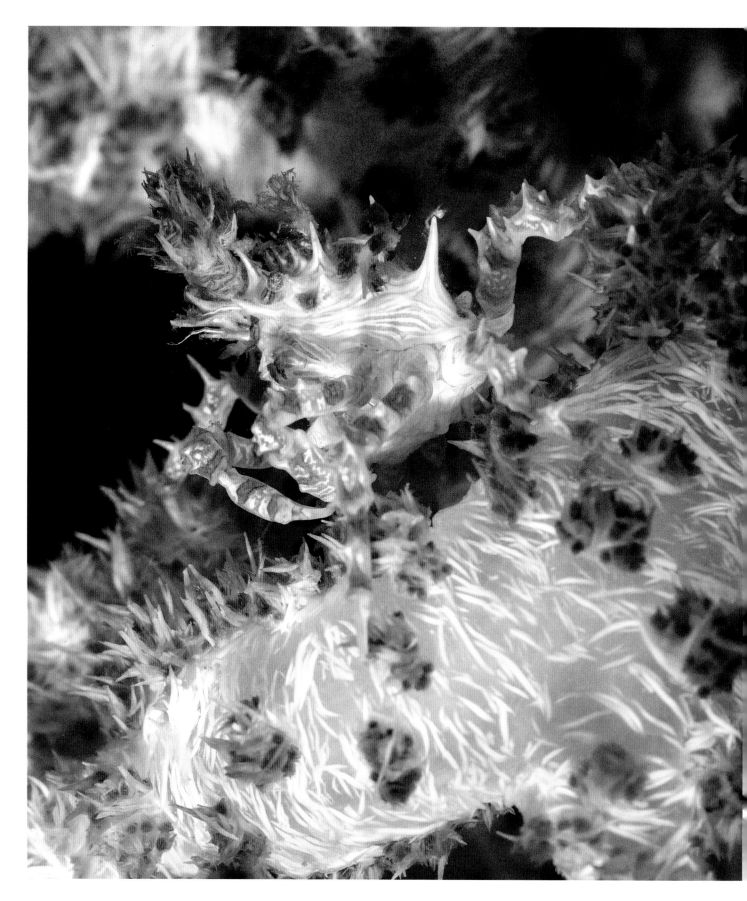

CANDY CRAB
Hoplophrys oatesii

FAMILY:	Epialtidae
OTHER NAMES:	Oates's soft coral crab, commensal soft coral crab, *Dendronephthya* crab
DISTRIBUTION:	Widespread tropical Indo-West Pacific
HABITAT:	Coral reefs, on species of *Dendronephthya*; to 90 m (300 ft) depth
FEEDING HABITS:	Plankton and detritus caught in the mucus of its host
NOTES:	Can be a range of colours to match those of its host
SIZE:	To about 20 mm (¾ in) carapace width

THE APTLY NAMED CANDY CRAB is indeed a confection of colour, although its prickly carapace and legs make it far from palatable. It is an excellent example of mimicry used by a range of crab species, but particularly embraced by spider crabs (superfamily Majoidea), which have numerous extraordinary mimics amongst their ranks. It is especially common amongst crabs that live on coral or other invertebrates, concealing themselves by matching the colours and patterns of their host.

SOFT CORAL DWELLER

Hoplophrys oatesii superbly mimicks the colours of the polyps and trunk of the distasteful and prickly soft coral (*Dendronephthya* species) upon which it always clings, often also attaching coral polyps to its carapace. Different crabs can assume a variety of colours to match those of its host, with variations of white, pink, yellow or red being typical. The tips of the walking legs are sharply clawed and have some small spines that help the crabs stay attached to the leathery surface of the soft coral. Not much is known of the crab's specific biology, but they apparently steal the plankton and detritus gathered as food by the *Dendronephthya*.

THE PERFECT 'SAFE HOUSE'
Prickly carnation corals (*Dendronephthya*) have stinging nematocysts and a tough, toxic body, so they are vulnerable to only a few species of marine snails and nudibranchs. With larger active predators keeping their distance, they are the ideal home for the small, camouflaged candy crab.

SUPERB DECORATOR CRAB
Camposcia retusa

THE SELF-ADORNMENT UNDERTAKEN by *Camposcia retusa* represents the *haute couture* of the crab world. While most crabs go to great lengths to keep themselves clean, decorator crabs deliberately cover every exposed upper surface in a variety of living marine organisms – algae, sponges, bryozoans, hydroids, anemones and ascidians (collectively termed 'epibionts'). This provides superb camouflage against the rock and coral bottoms, and allows them to graze for food, virtually invisible to predators. An added benefit may be that epibionts such as stinging hydroids and inedible toxic sponges will further deter attackers.

'PLANTING A GARDEN'

Decorator crabs often have slender, highly flexible claws that can reach behind their backs to 'plant' their chosen decorations, and the shells and legs of many species are covered in a meshwork of low, hooked bristles, rather like Velcro, which they use to attach pieces of sponge. They can shape the pieces quite precisely using their mouthparts, coating the ends with a glue secreted from a special gland. Once planted, the 'garden' continues to grow until the shape of the crab is completely obscured. When the crab grows enough, a moult is needed and all the hard work is shed with their shell, but then there is the opportunity to create a new masterpiece!

FAMILY:	Inachidae
OTHER NAMES:	Velcro crab
DISTRIBUTION:	Widespread through the tropical Indo-West Pacific
HABITAT:	Rocky and coral reefs; to 30 m (100 ft) depth
FEEDING HABITS:	Omnivorous; scavenger
NOTES:	Despite its camouflage, it is mainly active nocturnally
SIZE:	To about 40 mm (1 5/8 in) carapace length, but as much 13 cm (5 in), including legs

DRIVEN TO BE INDIVIDUAL
Arguably the most overdressed decorator crabs of the reef, in full regalia they can be almost impossible to distinguish – despite the fact that they are relatively large crabs. They also plant themselves with such variety that no two ever look the same.

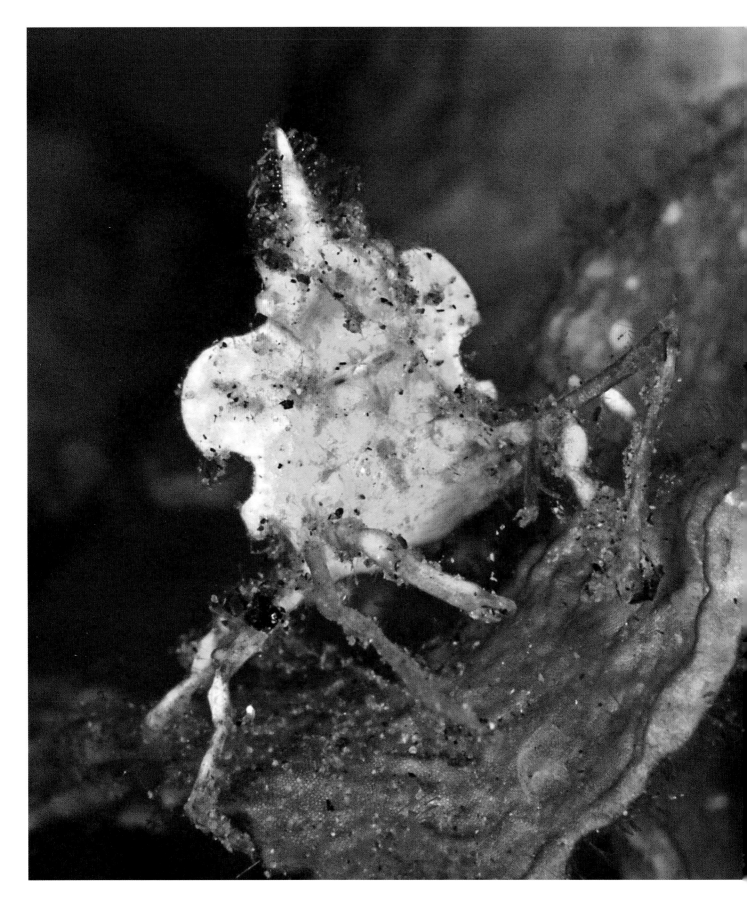

ARROWHEAD CRAB
Huenia heraldica

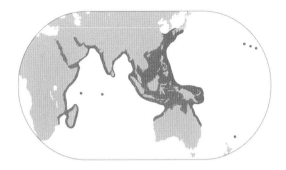

FAMILY:	Epialtidae
OTHER NAMES:	Halimeda crab
DISTRIBUTION:	Widespread Indo-West Pacific: East Africa to Japan, Australia, Hawaii and Kermadec Islands
HABITAT:	Shallow water reefs, 0–36 m (120 ft)
FEEDING HABITS:	Omnivorous grazer
NOTES:	Carapace shape changes dramatically between males and females
SIZE:	To about 25 mm (1 in) carapace width

ARROWHEAD CRABS BELONG to the family Epialtidae, which is one of several spider crab families grouped together in the superfamily Majoidea – one of the oldest branches of the crab evolutionary tree. Majoids are extremely variable in appearance, but have a characteristic projecting frontal rostrum between the eyes, and tend to be relatively slow-moving bottom dwellers. This lifestyle has led many to adopt some form of camouflage, and the decorator crabs are well-known and spectacular exponents of the art (see page 168).

SHAPE SHIFTER

Some spider crabs, however, take a different approach by evolving fundamental changes in the shapes of their carapace and legs to match the substrate on which they live. Species of *Huenia* are particularly adept at this, and *Huenia heraldica* perfectly mimics the shape and colour of large calcified segments of *Halimeda* algae (green or brown). The crab also plants small alga fronds on its rostrum to complete the illusion. The carapace shape varies according to sex and age, too – sometimes quite dramatically – further confusing predators. An ability to change colour to match the colour of the alga is the last trick up its sleeve. Researchers have invoked colour changes in several epialtid crab species by simply feeding them algae of different colours.

VIVE LA DIFFÉRENCE

The bodies of male arrowhead crabs look just like long, triangular arrowheads, but in mature females the front half of the carapace has wide, flattened flanges that transform its shape to resemble the fronds of the green *Halimeda* alga on which it lives.

STALK-EYED SHORE CRAB
Tmethypocoelis ceratophora

EVERYDAY, MANY TINY GEMS ARE OVERLOOKED. Living on the muddy sand banks of mangrove estuaries in central China and Taiwan is a remarkable little crab, whose almost unpronounceable scientific name is much bigger than it is! Derived from Greek the genus name simply means these crabs have a divided margin below the eye, a feature that was considered to split them away from their other known relatives back when it was first described in 1897. This is combined with *ceratophora*, which means 'stalked eye'.

DANCE PARTNERS

Like their larger and more colourful companions, the fiddler crabs, females have only small, inconspicuous claws, used for scouring the mud for food and for personal grooming. But unlike fiddlers, male *Tmethypocoelis* have both claws equally enlarged, and they put these to good use performing ritualistic dance moves to attract partners (and to aggressively push and shove rival males). Each species has its own special performance. *Tmethypocoelis ceratophora* begins with the chelipeds held in front of the face, then extends them sideways and upwards before refolding them again (see page 145). When dotillid crabs such as these are in large numbers, they will often synchronize with each other, sending a tiny Mexican wave rippling across the surface of the mud.

FAMILY:	Dotillidae
OTHER NAMES:	Buddhist crab
DISTRIBUTION:	East Asia: South China and Taiwan
HABITAT:	Estuarine; sandy to gritty mud near mangroves
FEEDING HABITS:	Deposit feeder on the mud's surface
NOTES:	Currently four described species of *Tmethypocoelis* stretching from southern Japan south to northern Australia, although each with limited distribution
SIZE:	To about 10 mm (3/8 in) carapace width

ATTRACTING A PARTNER
Male crabs perform elaborate displays to attract females. This handsome male has been caught mid-performance, with his chelipeds in the upright position.

THICK-LEGGED FIDDLER CRAB
Paraleptuca crassipes

FAMILY:	Ocypodidae
OTHER NAMES:	Estuary fiddler crab
DISTRIBUTION:	Indo-West Pacific: from southern Japan south to Indonesia, New Guinea, eastern Australia, and the islands of Melanesia and Micronesia
HABITAT:	Intertidal in estuaries and mangroves
FEEDING HABITS:	Detritovore
NOTES:	Originally described from the Philippines, recent genetic research suggests there may be several similar species confused together across the range
SIZE:	Up to 22 mm (7/8 in) carapace width

FIDDLER CRABS ARE SOME OF THE MOST SPECTACULAR of the intertidal crabs. There are around 105 species in 11 genera spread across the tropical and subtropical coasts of the world, typically found on soft substrates around mangroves and saltmarshes, often penetrating into brackish estuarine waters.

A LOVER AND A FIGHTER

Male fiddlers are remarkable in having one enormously enlarged, often brightly coloured claw. They use this as a weapon to fight off other males (typically pushing contests), or more importantly, to lure females for mating. Each species has its own colour pattern and its own sequence of waving movements to which only the females of its own species will respond. Watching these crabs on the mud flat, some whimsical observers have been reminded of a violinist's bow rising and falling, giving rise to their common name. Hundreds of crab courtiers all flashing their claws together in the sun can be a spectacular display. *Uca* species feed by scraping the surface of the mud and sand for microscopic organisms, so it is fascinating that males are left with only one useful food-gathering claw.

CRYPTIC SPECIES

This beautiful fiddler crab from New Caledonia is currently identified as *Paraleptuca crassipes*, but genetic studies are suggesting that all may not be as it seems. Further research is needed to understand the speciation of these crabs in the western Pacific region.

GARFUNKEL'S CRAB
Danarma garfunkel

GARFUNKEL'S CRAB IS ONE OF A GROUP of five similar-looking specialist Indo-West Pacific species that make their home in the harsh, high, supratidal zone, living in crevices and holes amongst eroded limestone and other rocks. Each species seems to be narrowly confined to small tropical island groups, such as Hawai'i, Guam, Taiwan (two species), the Ryukyu Islands, and Christmas Island in the Indian Ocean. Such hot, dry, desolate habitats mean they are mostly restricted to night-time activity, when the risk of dehydration is lessened; only during the monsoonal rainy season might they occasionally be found wandering freely. To get water they must rely on sea spray carried high onto the rocks by onshore winds, rain caught in the cracks and pockmarks in the rocks, and the dew fall of hot humid nights.

MOTHERHOOD HAS ITS CHALLENGES

Like other semi-terrestrial crabs, they must return to the sea to release their eggs. Females typically carry 5,000 to 7,000 eggs, and have been found to spawn any time from November to April. The small crabs make a perilous journey down from the high limestone terraces surrounding the island to deposit their eggs at low tide, in rock pools on the fringing reef flat, but in some situations they will also spawn from the edge of the sea cliffs themselves.

FAMILY:	Sesarmidae
DISTRIBUTION:	Indigenous to Christmas Island, northeast Indian Ocean
HABITAT:	Coastal, supralittoral to 50 m (165 ft) above sea level, but most common in the spray zone; in crevices and holes in coastal limestone
FEEDING HABITS:	Presumed omnivorous; probably a nocturnal predator of micro-snails and other small invertebrates
NOTES:	Although first discovered in 1934, this species remained wrongly identified until being recognized as a separate novel species in 2013
SIZE:	To about 25 mm (1 in) carapace width

BRIGHT EYES

This species was named for Art Garfunkel, who performed the beautiful song 'Bright Eyes' from the 1978 animated film *Watership Down*. Like the rabbit to which the song alludes, this crab is also nocturnal, lives in holes, and is remarkable for its bright, shining eyes.

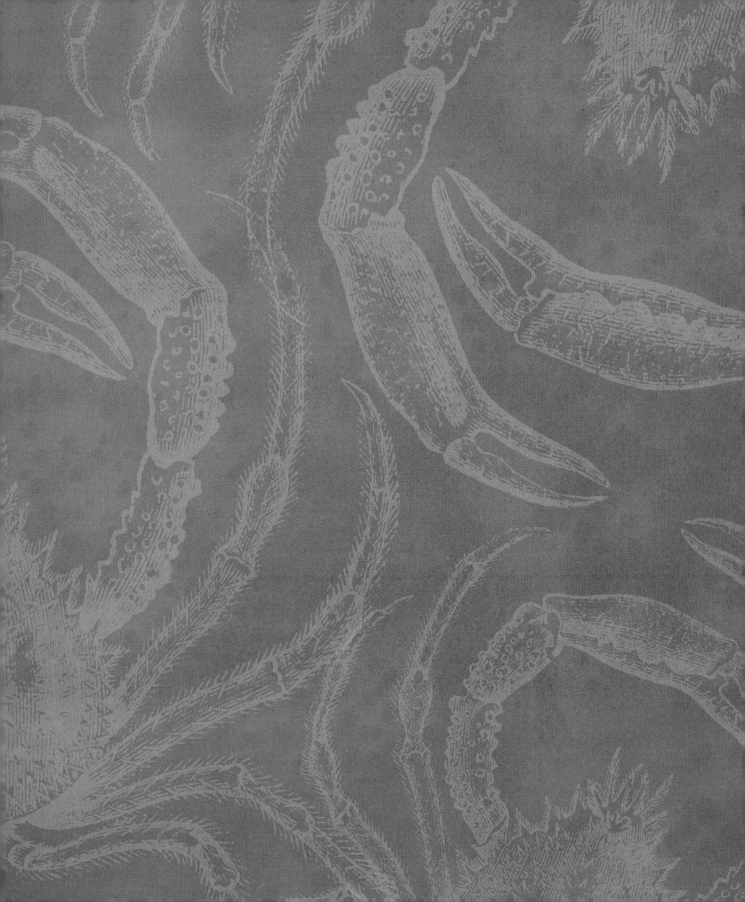

5
EXPLOITATION AND CONSERVATION

CRABS AS FOOD

CRABS HAVE BEEN PRIZED AS FOOD since ancient times, and probably since the first humans stood erect and left the forests to form coastal settlements. Today there are multimillion-dollar industries based on crab fishing and aquaculture, with crabs exported to markets across the globe. Crab food festivals are staged in small coastal towns worldwide to celebrate the bounty from the sea, and to promote tourism. Around 1.5 million tonnes of true crabs are fished annually, which is about 20 per cent of all decapod crustaceans caught and farmed globally. While there is a large regulated industry, there are also millions of recreational fishers, and indigenous communities continue to eat crabs as part of their traditional diet. Most of the crabs forming the commercial industry belong to about 14 brachyuran families, and a single family of anomurans (Lithodidae – stone or king crabs; see page 186).

BELOW: The snow crab (*Chionoecetes opilio*) is mostly caught by trapping, and only males are kept. Native to the North Pacific it has now been introduced to the Barents Sea in the northeast Atlantic.

CULTURAL FOOD TRADITIONS

Many poorer communities, and those peoples still living a more traditional lifestyle, eat a much wider variety of crabs than those considered commercial – even quite small crabs, if they can collect them in large enough numbers. These include many intertidal ocypodoid and grapsoid species such as dotillid soldier crabs (*Dotilla* species), fiddler crabs (*Uca* species), and periscope or sentinel crabs (*Macrophtalmus* species), as well as several types of larger 'vinegar crabs' (Sesarmidae, Grapsidae). Indigenous peoples of several Pacific islands also consume a number of medium-sized reef crabs in the families Xanthidae, Oziidae, Menippidae and Eriphiidae. Several species of freshwater crab of the families Potamidae and Gecarcinucidae are consumed in many parts of Africa, southern Europe, Southeast Asia and Indochina. Blue land crabs (*Cardisoma guanhumi*) and purple land crabs (*Gecarcinus ruricola*), both in the family Gecarcinidae, are often harvested by locals around the Caribbean; the latter being renowned for occurring in very large numbers during their annual breeding migrations.

The mud- and mangrove-dwelling swamp ghost crab (*Ucides cordatus*) of the central west Americas is a favourite in many countries in the region. This species is especially important to the local fisheries in Suriname and French Guyana, where it is traditionally collected by hand from burrows and marketed fresh or cooked.

TOP LEFT: One of the larger sentinel crab species, *Macrophthalmus abbreviatus*.

TOP RIGHT: A Southeast Asian mangrove vinegar crab (*Episesarma versicolor*).

CENTRE: A swamp ghost crab walking across a muddy sand beach.

RIGHT: Fresh swamp ghost crabs ready for market in Brazil (locally known as *caranguejo uçá*).

FAR LEFT: *Metaplax gocongensis* is a favourite seasonal delicacy for the people of the Mekong Delta in southern Vietnam.

LEFT: Tiny adult swimming crabs (*Charybdis brevispinosa*) are trawled from the South China Sea, deep-fried, and sold as crunchy snacks in China.

The robber or coconut crab (*Birgus latro*; see page 44) is widely eaten on the islands of the western Pacific, and has sometimes formed the basis of local industries. However, this very slow-growing species, which can live to 60 years old, is now increasingly rare from over-exploitation and can be considered vulnerable or endangered in many parts of its range.

An interesting story concerns the intertidal estuarine species *Metaplax gocongensis* (Varunidae) of South Vietnam. This moderate-sized crab (large males reach 30 mm [1⅛ in] across the shell) is native to the Mekong Delta. A gregarious species, it digs simple burrows about half a metre deep in intertidal silty mud. The fifth day of the fifth month of each lunar year (typically early June) has historically marked the beginning of heavy rains in the region, and great numbers of this species have emerged from their burrows to head into nearby wet vegetation, freshwater pools or water-filled tracks to undergo a synchronized mass moulting. On this one day of the year, local people have traditionally collected large quantities of these soft-shell crabs for making a regional dish named *mam cong lot* (salty moulted-crab). The carapace, abdomen, stomach and gills are discarded, and the rest of the crab is preserved in brine. This special dish is highly regarded by older people of the Gocong Province where this species has flourished, but also by some gourmets in Saigon. Since about 1995, however, this extraordinary annual event has been seriously affected by climatic changes, with extended and unpredictable wet or dry periods leading to poor or failed moulting events. The effects of climate change could well have a devastating impact on the future of this unusual species.

RIGHT: Live orange mud crabs (*Scylla olivacea*) trussed up safely for market sale. This is one of four similar species commercially marketed in the Indo-West Pacific region.

THE MAJOR COMMERCIAL FISHERIES

The largest true-crab fishery in the world, with a catch of around 350,000 tonnes a year, is based on an East Asian swimming crab, the Gazami crab (*Portunus trituberculatus*), also known as the Japanese blue crab or horse crab. Other portunid crabs, such as the blue-swimmer or flower crabs (*Portunus pelagicus* and related species), giant mud crabs (*Scylla serrata* and related species) and blue crabs (*Callinectes sapidus*), also constitute major tropical fisheries. Temperate-water species have important fisheries, too. The snow crab (*Chionoecetes opilio*) is particularly important in Canada, where 92,850 tonnes, worth CAN$430 million, were landed in 2012 (see also page 180). Edible or brown crabs (*Cancer pagurus*) and Dungeness crab (*Metacarcinus magister*) each yield at least 20,000 tonnes annually.

Portunus trituberculatus

ABOVE AND BELOW: The Gazami crab (above), the Asian flower crab (below) and several other closely related Indo-West Pacific species of the genus *Portunus,* are the basis of important commercial fisheries.

CRABS AS FOOD

'BOUTIQUE' FISHERIES

The coastal North Atlantic common spider crab (*Maja brachydactyla*), and the closely related spiny spider crab (*Maja squinado*) found only in the Mediterranean, are together a favourite European delicacy, with over 6,000 tonnes brought ashore annually – more than 70 per cent of it from off the coast of France (see also page 208). Across the world, the tiger crab (*Orithyia sinica*, page 206) is found only along the coast of mainland Asia, from South Korea to Hong Kong. It is fished on a small scale (usually with tangle nets in rocky areas), but commands high prices.

The Florida stone crab (*Menippe mercenaria*) and its close relative the Gulf stone crab (*Menippe adina*) occur in the western North Atlantic coast of the Americas, from Connecticut to Belize, the Gulf of Mexico, Cuba and Bahamas. The Florida stone crab is usually fished near jetties, oyster reefs or other rocky areas. It is an unusual fishery because harvesting is accomplished by removing one or both large meaty claws, and returning the live crabs to the ocean with the intention that they will regenerate their lost limbs. Even so, mortality rates can be up to 60 per cent, and it has been estimated that only about 20 per cent of claws are regrown.

Crabs of the family Geryonidae are typically deep-sea inhabitants living on muddy and sandy-mud substrates between 50 and 2,800 m (165–9,200 ft) depth, where temperatures range from 4 to 12°C (39–54°F). There are six species commonly fished. Of those, *Chaceon quinquedens* is the basis of a sustainable fishery along the Atlantic coasts of Canada and the United States, where a fishery for *C. fenneri* also takes place; and *C. notialis* has been exploited in Uruguayan waters since the 1990s. In Australia there is a small but important fishery for the crystal crab (*Chaceon albus*). Deep-water crab fisheries need to be carefully managed because these crabs grow slowly and often have long lives (see below).

Another deep-sea species, the giant crab (*Pseudocarcinus gigas*) is the basis for a comparatively small annual catch (only 50–60 tonnes), but it is worth more than $2 million to the Australian economy each year. Much of the catch is exported live to the Asian restaurant trade. The giant crab is considered to consist of a single biological stock across southern Australia, however the fishery is managed separately from state to state. Although it has only been commercially targeted since the early 1990s, the fishery is now considered to be fully exploited and has moved from being open access to being strictly controlled by individual licences and quotas. Even so, stock is currently considered depleted in Tasmania.

Spanner crabs (*Ranina ranina*) are widespread throughout the Indo-West Pacific region, in sandy open bottoms at depths of 10–100 m (33–330 ft). While they are commercially exploited in a number of countries, the most significant commercial industry is along the mid-eastern coast of Australia, with an annual catch of around 3,600 tonnes. Much care and research has gone into ensuring the fishery is well managed, and it was one of the first to meet government criteria for ecologically sustainable fisheries (see page 204).

OPPOSITE LEFT: A mating pair of golden deep-sea crabs, (*Chaceon fenneri*). Like other species of *Chaceon*, this crab is the subject of a small fishery in the tropical West Atlantic.

OPPOSITE RIGHT: Florida stone crabs are fished near jetties and rocky reefs. Typically only the meaty claws are removed, and the live crabs are released, but mortality rates are high.

ABOVE: The small annual catch of the Australian giant crab is mostly exported live to the Asian restaurant trade. The fishery is now fully exploited and very strictly controlled.

KING CRABS

King crabs are 'crab-like' rather than true crabs. They are most abundant in cold temperate waters, although many species occur in the deeper, colder waters of tropical oceans. Around 121 species belonging to 10 genera are known, of which about 12 species are fished to a greater or lesser extent, and three are heavily exploited. The red king crab (*Paralithodes camtschaticus*; see page 202), the blue king crab (*Paralithodes platypus*) and the golden king crab (*Lithodes aequispinus*) are each the basis of important fisheries in the north Pacific, from off the coasts of Japan and Russia, through the Bering Sea and off the coasts of southern Alaska and northeastern Canada. Unfortunately, some areas have had to be closed to fishing in recent years as populations have become unstable due to overfishing.

ABOVE: A horsehair crab (*Erimacrus isenbeckii*) on ice, awaiting sale. Its meat is particularly prized in Japanese cuisine (see also page 211).

OPPOSITE: Baird's tanner crab (*Chionoecetes bairdi*) is the subject of an important fishery in the North Pacific. Here, the male crab is seen guarding the female prior to mating.

ABOVE: The red deep-sea crab (*Chaceon quinquedens*) eating a meal of molluscan eggs. It is the basis of a sustainable fishery in deep waters off the Atlantic coasts of Canada and the US.

The red king crab is a very large species, and often comprises 90 per cent of the annual king crab harvest. It lives in shallower depths to about 50 m (165 ft) on muddy or sandy substrates. Blue king crabs mostly prefer deeper waters with cobble, gravel and rock bottoms. Golden king crabs can be extremely abundant, but are smaller in size, averaging 2.3 to 5.5 kg (5–12 lb), and prefer to live on steep-sided slopes in depths that can exceed 550 m (1,800 ft). In what some consider an ecological disaster, both red king crabs and snow crabs were intentionally introduced to the Barents Sea and European Arctic waters, and both are spreading rapidly southwards, but fisheries for them are also expanding in this new region.

OVER-EXPLOITATION

Not all crab fisheries have been well managed in the past. This has been the result of commercial greed, highly efficient modern fishing technologies, and because there has been too little understanding of the biology and lifestyles of the crabs concerned. Sustainable fisheries must be based on knowledge, and catch limits must be strictly controlled. This is especially important for many deep-sea species, which are often very long-lived, grow very slowly, take a long time to become sexually mature, and may not reproduce every year. Such species typically have low annual juvenile recruitment and therefore slow recovery when intensely exploited. Removing these species from their deep-water food webs also has significant and far-reaching ecological consequences.

An uncontrolled fishery of deep-water crabs (and other animals) typically follows a pattern of rapid annual increases in catches, followed by an abrupt decline, marking the end of their commercial viability and potentially threatening the existence of the species itself. An example of this is the royal crab (*Chaceon ramosae*), which occurs from 350 to 1,200 m (1,150–4,000 ft) deep off southeastern and southern Brazil. From 1999 to 2009, around 585 tonnes per year of this species was taken, reaching a high of 1,742 tonnes in 2004. The population finally crashed – estimates indicate that it was nearly halved in a few short years – and the fishery was forced to close. Fortunately, it continues to survive in greater depths and in areas that were outside the range of the industrial fishing fleet.

A similar tale can be told for the Australian champagne crab (*Hypothalassia acerba*), a large spiny crab with a carapace width up to 15 cm (6 in). It is indigenous to the continental shelf off southwestern Australia, from about 100 to 540 m (330–1,770 ft) depth. Its fishery potential was first recognized in 1966, but only escalated to a serious commercial industry through the 1990s, reaching an annual catch of about 45 tonnes before stocks dropped dramatically. The crabs are caught using baited traps placed on the open muddy bottoms. Very small catches are still made, and primarily sent live to the lucrative markets of China and Singapore, where it is highly valued. Management is now focused on safeguarding breeding stocks; females with eggs must be returned to the water, and males must reach a minimum size before they can be retained.

By contrast, the northwest Atlantic deep-sea red crab (*Chaceon quinquedens*) is a good example of a small, sustainable fishery that operates along the US coast, from North Carolina to the Canadian border, in depths of 400 to 800 m (1,300–2,600 ft). A moderate-sized species (fished crabs typically weigh less than a kilogram), it has an annual quota of only about 1.8 tonnes, and females are all returned to the water to maintain breeding stock. The crabs are caught in specially designed large conical traps that are left in the water for 24 hours at a time.

AQUACULTURE

Farming of crabs currently yields much less than the wild catch, but it is still a major endeavour, with about a million tonnes of crabs produced annually. China is responsible for around 95 per cent of the world's production, and more than 700,000 tonnes comes from just one species, the Chinese mitten crab (*Eriocheir sinensis*). The aquaculture of these crabs developed rapidly during the 1990s, with new methods for making larval rearing practical, and enabling farms to be established across China. Most crab 'seed' (megalopae, the first bottom-dwelling settlement stage, see page 151) are produced in hatcheries, mainly located in the Yangtze River Delta in southern China, then transported to other parts of the country for grow-out. The numbers are astonishing – in 2003, 522,893 kg of live megalopae were produced. These nascent crabs are so tiny that a kilogram equates to about 140,000 individuals, meaning that the annual stocking production exceeds 7.3×10^{10} (many billions of baby crabs!). The sale of mitten crabs is seasonal, starting in autumn in Shanghai and eastern China, and the roe of female crabs is particularly prized. In Chinese culture, the meat is believed to have a 'cooling' (*yin*) effect.

Several species belonging to another family, the Portunidae (swimming crabs), are also cultured. These include four types of giant mud crab, *Scylla* species, and the flower crab (*Portunus pelagicus*). While the market for mitten crabs is focused on China, there is a much wider Indo-Pacific market for portunid crabs. Production was historically based on collecting juvenile wild crabs and stocking them at low densities in pens within coastal mangrove swamps, where they would grow to saleable size. But in recent times the techniques needed to raise larvae through to juvenile crabs have been developed, and this has meant mass production by the typical sequence of hatchery, nursery and grow-out phases, much as occurs with mitten crabs. A serious drawback with portunids, however, is their aggressive predatory behaviour – meaning that stocking densities must be kept low to reduce cannibalism.

Aquaculture can have a downside, though. Over the last 40 years, Myanmar has lost more than 1 million hectares (about 2.5 million acres) of mangroves to aquaculture of fish and crustaceans, as well as to other coastal agriculture. Such devastation is widespread throughout tropical Southeast Asia. Mangroves protect coastlines from erosion during storms and in the face of rising sea levels, and are vital for maintaining coastal ecology and sustainable fisheries stocks. Active programmes of coastal rehabilitation and mangrove reafforestation are now finally being increasingly undertaken.

OPPOSITE: This spiny deep-water crab, *Hypothalassia armata* from the western Pacific, is almost identical in appearance to – and long confused with – the Australian champagne crab. Both are found at depths of over 500 m (1,640 ft).

ABOVE: A crab-fattening pond in Myanmar's low-lying Ayeyarwady Delta. Intensive coastal agriculture and aquaculture mean that now only 16 per cent of the country's original mangrove cover remains.

CRABS AND DISEASE

SOME CRABS HAVE CAUSED humans serious health problems, in rather unexpected ways. In Southeast Asia, Africa and the Neotropics, freshwater crabs can be medically important as intermediate hosts of lung flukes (mainly of the genus *Paragonimus*), causing the condition known as paragonimiasis in humans. Paragonimiasis is a food-borne zoonosis (a disease that can be transmitted to humans from animals), and the fact that more than 20 million people are infected worldwide by one of the 15 species of lung flukes indicates how widely spread is the consumption of freshwater crabs. Crabs become infected by exposure to infectious cercariae in the water, or by eating infected snails, and in turn people become infected by eating raw or poorly cooked crabmeat (especially from treats such as 'drunken crab', where live animals are marinated in wine before eating).

GEOGRAPHIC DISTRIBUTION OF PARAGONIMIASIS

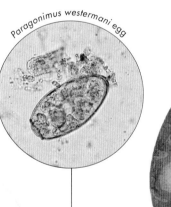

ABOVE: The lung fluke (*Paragonimus westermani*) – both as an egg and an adult – for which crabs such as *Eriocheir* species are an intermediate host.

LEFT: The popular Chinese mitten crab is a Chinese delicacy that, if poorly cooked, can transfer the *Paragonimus* lung fluke to humans.

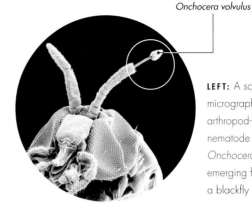

LEFT: A scanning electron micrograph showing the arthropod-borne filarial nematode (roundworm) *Onchocerca volvulus* emerging from the top of a blackfly larva's antenna.

LUNG FLUKES AND THE ANCIENT EGYPTIANS

The connection between crabs and lung flukes appears to have been understood since ancient Egyptian times. According to the Egyptian legend (as later related by the Greek historian Plutarch, who lived from AD 46 to AD 120), the ruler of Egypt, Osiris, was killed by his brother Set, a rival to the throne. Set cut his brother's body into many pieces that were then distributed all over the country. Osiris' distraught wife, Isis, collected all the pieces in order to restore Osiris to life, but his penis had already been eaten by the river crabs. For this sin, Isis cursed the crabs by making them poisonous, and anyone eating them would suffer the punishment of the gods and become ill. So even in ancient times it appears that fear of divine punishment put an end to the consumption of these crabs in North Africa, and protected against a deadly parasitic infection.

In Asia, two commonly eaten species of mitten crabs, *Eriocheir sinensis* and *Eriocheir japonica*, also act as important vectors, with infection rates in crabs reaching almost 60 per cent in some areas.

Some freshwater crabs of the family Potamonautidae that live in the highland streams of Kenya, Uganda and Cameroon play host to the developing larvae of biting blackflies (*Simulium* species). Blackfly larvae inhabit fast-flowing streams with adjacent dense forest canopies, and use the carapace of the crab host as a site of attachment. These larvae are the vectors of the nematode parasite *Onchocerca volvulus*, the cause of hundreds of thousands of human cases of onchocerciasis (river blindness) in Africa, the next most important cause of blindness worldwide after trachoma.

ABOVE: Species of *Potamonautes* can play host to blackfly larvae – vectors for river blindness. However, this species, *Potamonautes lirrangensis*, is widespread across central and eastern Africa, and is the basis of a small commercial fishery in Malawi.

CRABS AND DISEASE 191

TOXIC CRABS

Many people are surprised to learn that some crab species can be highly toxic if eaten, and there have been a significant number of recorded deaths throughout the Asia-Pacific region, including the Philippines, Japan, East Timor, Palau, Vanuatu, Fiji and Mauritius. Over 50 species have so far been found to harbour potentially lethal levels of toxin. Crabs do not make their own unique toxins, but instead sequester into their flesh a variety of toxins from their environment. Five groups of toxins that have so far been identified in crabs are shown below.

ABOVE: An inhabitant of Indo-West Pacific reefs, the shawl crab (*Atergatis floridus*), named for its lacy patterning, is almost always highly toxic.

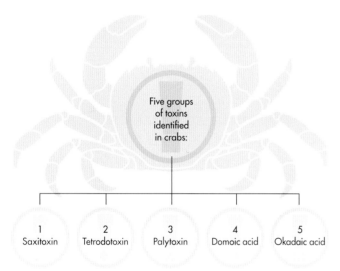

TOXINS IN CRABS

Five groups of toxins identified in crabs:
1. Saxitoxin
2. Tetrodotoxin
3. Palytoxin
4. Domoic acid
5. Okadaic acid

Saxitoxin ('paralytic shellfish poison') is typically produced by toxic 'algal' dinoflagellate and cyanobacterial blooms, and is best known from poisonings caused by eating filter-feeding molluscs such as oysters, clams and scallops, making it a serious health threat worldwide. It is the most common crab toxin, found in over 90 per cent of poisonous brachyurans. The common north American mole or sand crab (*Emerita analoga*; page 26), while not typically eaten by humans, is efficient at accumulating saxitoxin and its derivatives. Routine screening for saxitoxin already occurs for several commercially important crabs that potentially feed on contaminated shellfish, such as the Dungeness crab (*Cancer magister*), tanner or snow crabs (*Chionoecetes bairdi* and *C. opilio*) and king crabs (*Paralithodes* species and *Lithodes aequispina*), and fisheries have very occasionally been closed.

Tetrodotoxin is the infamous 'fugu' toxin found in pufferfish. Like saxitoxin, it is produced by bacteria, but the pathway to its presence in crabs is unknown. It is the second most common crab toxin, occurring in 20 per cent of toxic species. Saxitoxin and tetrodotoxin are neurotoxins that block the sodium channels that carry messages between the brain and muscles. Sensation is lost as the poison takes effect, rapidly followed by muscle paralysis that extends to the diaphragm and other muscles that enable breathing. Tetrodotoxin is around 1,200 times more deadly than cyanide!

Palytoxin, domoic acid and okadaic acid all have a low incidence in crabs. Palytoxin is a complex natural product that attacks the 'sodium–potassium pump' found in cell membranes. It has been discovered in an array of organisms but is best known in a mat-forming cnidarian (*Palythoa*) common in tropical and subtropical reefs – hence its name.

Domoic and okadaic acids, like saxitoxin, are typically found in mussels and clams that have consumed large amounts of toxic microalgae such as diatoms (*Pseudonitzschia*) and planktonic dinoflagellates. Domoic acid can cause headache, dizziness, seizures and even death. It is often referred to as 'amnesic shellfish poisoning' because short-term memory loss is common. It has been identified in the Dungeness crab, tanner crab and red rock crab (*Cancer productus*) from the north coasts of North America. Okadaic acid causes 'diarrhetic shellfish poisoning', commonly resulting in diarrhoea and vomiting. Cases have occurred following consumption of green crab (*Carcinus maenas*) in Portugal, and brown crab (*Cancer pagurus*) in Norway.

HOW DO CRABS BECOME POISONOUS?

Most toxic crabs are omnivorous or scavenging species for whom the specific source of toxins remains a mystery. For many, toxicity may vary from time to time, and place to place, but some widespread species, like the shawl crab (*Atergatis floridus*) and devil crab (*Zosimus aeneus*), are almost routinely toxic. They both belong to a group called the 'black-fingered reef crabs' (superfamily *Xanthoidea*). Some xanthoid crabs have even been found to contain saxitoxin, tetrodotoxin and palytoxin in potent combination.

It seems increasingly clear that all crab toxins are originally derived from either bacteria or phycotoxins (a diverse group of chemical compounds produced by microalgae). Most phycotoxins are produced by dinoflagellates, diatoms and cyanobacteria. Macro-algae are important in the diet of many crabs, but are rarely toxic in themselves. They do, however, often harbour colonies of toxic dinoflagellates (such as *Ostreopsis* species) that are then also inadvertently eaten. Such toxins then pass up the food chain. Crabs have evolved a high tolerance to the presence of these toxins in their flesh. Shawl and devil crabs alone may harbour several tonnes of saxitoxin – enough to kill hundreds of millions of people. The poisonous flesh of these large, often colourful species presumably has the added benefit of deterring predators from eating them. The message is simple: avoid eating tropical reef crabs at all costs, no matter how large and appealing – they could be your last meal!

LEFT: Species of the genus *Demania* received a lot of study in the 1970s after a fatal poisoning in the Philippines. This species, *Demania splendida*, has not been tested, but at least four other species have been shown to contain saxitoxin, tetrodotoxin or palytoxin.

CRABS BEHAVING BADLY

INVASIVE MARINE SPECIES (species that become established in new areas away from their normal range) have become an increasingly significant problem worldwide. This is due to a number of factors, but the most important has been the dramatic growth in international shipping. Adult crabs can often hitch a ride by living in the fouling communities on the bottom of both commercial and recreational ships of all sizes, but their larvae can also be transported across oceans by surviving in the ballast water that empty cargo ships take aboard to improve stability at sea. This water, and its living cargo, is then discharged at the destination port during unloading. Invasive species can also be spread through man-made canals, importation by the aquarium trade, intentional release to establish a new fishery, and escape into the wild from coastal aquaculture facilities.

Opened in 1869, the Suez Canal was built to connect the Red and Mediterranean Seas, and it has long been a favourite route for crabs to invade new territories. The Red Sea is about 1.2 m (4 ft) higher than the Mediterranean, so there is a natural net flow of seawater to the north through the canal. This, combined with the lack of locks and any intervening freshwater passages, means there are no impediments to prevent migration. To date, about 30 Indian Ocean crab species have made the journey to colonize the Mediterranean, accounting for around 40 per cent of known invasive crabs worldwide. This phenomenon has been named 'Lessepsian migration', after Ferdinand de Lesseps, the French official in charge of building the canal.

It is rare for crab invaders to survive at their new locality initially, but if conditions are right, and they can establish breeding populations, then local coastal currents may quickly transport their larvae far and wide. Their new habitat range is ultimately defined by their physiological limitations,

LEFT: The rosy egg crab (*Atergatis roseus*), is one of over 30 successful Lessepsian brachyuran immigrants. Arriving in Israel in 1961 via the Suez Canal, it is now well established in the eastern Mediterranean.

OPPOSITE: The European green crab has invaded temperate shores throughout the world. This example was photographed in Tasmania, Australia.

especially temperature tolerance, but also by biotic factors such as competition with, and predation by, other species. The most successful invaders breed in large numbers and are often aggressive predators or omnivores that can overpower local crabs. Invasive brachyurans can exert devastating ecological and economic impacts on their new habitats, and most countries regard these alien species as serious threats to native ecosystems. At least 73 species of alien brachyurans and crab-like anomurans have now been found, and of these, 52 species have become established.

EUROPEAN GREEN CRAB

Carcinus maenas (family Cacinidae) is the most widespread of invasive brachyurans, and has taken advantage of both ship hull fouling and ballast water to aid its travels. Native to the east Atlantic coasts of Europe and North Africa, it is now found in the temperate waters of South Africa, Australia and Japan, and along both coasts of North America, where it was first recorded near Massachusetts as long ago as 1817. Its intolerance of warmer subtropical waters seems to be its Achilles heel, restricting it from world domination. Despite being established for more than 100 years in southeastern Australia, the European green crab has not spread more broadly across southern temperate Australia, probably because the steep wave-battered cliffs of the Great Australian Bight are not suitable habitat. This crab loves to eat shellfish, so it is a particular threat to farmed mussels and oysters. Ironically, in its native Europe, a 20-year Dutch study has chronicled the demise of green crab populations because of competition and predation by the similarly invasive Pacific Asian shore crabs *Hemigrapsus sanguineus* and *Hemigrapsus takanoi*. *Hemigrapsus sanguineus* was first recorded in the Netherlands in 1999, and both species are spreading south down the coast of France.

LEFT: The red king crab was introduced to the North Atlantic by Soviet-era scientists. Although now commercially fished in Norway, it is causing serious ecological damage.

OPPOSITE: The Pacific snow crab was reported in the Barents Sea in 1996, larvae likely arriving in ballast water. It has since sparked an international incident over fishing rights!

CHINESE MITTEN CRAB

Eriocheir sinensis (family Varunidae) has arguably had the greatest impact of any invasive crab. Ballast water seems to be a primary mode of transport to new localities. It has few natural enemies, although fish, frogs and birds do eat smaller crabs. Vast numbers of crabs clog filters, and their extensive burrowing along riverbanks makes them vulnerable to collapses that lead to flooding. It was first found in the wild in Germany in 1912, but it quickly spread through the European canal network to establish itself across most of Europe and the Mediterranean, the Baltic and Black Seas, and England. More recently it has been introduced to the Pacific coast of North America, and to Chesapeake and Delaware Bays on the Atlantic coast – it is feared it will also become established in the Great Lakes. In the last 15 years it has also been reported in Ireland, Iran and Iraq, and Singapore.

RED KING CRAB

Soviet scientists deliberately introduced the red king crab into the northeastern Atlantic in 1960; it subsequently migrated south and become well established in Norway (see page 202). Like many invasive species, it is a generalist predator, eating anything and everything, from small invertebrates to large echinoderms and bivalves. While it may be a commercial boon to the Norwegian economy, the ecological damage it wreaks should not be underestimated. It is a large, fast-moving species that ranges over a broad depth range, so it can make rapid and significant changes to the community composition of the coastal shelf. Its consumption of larger bivalves and echinoderms has already led to lower diversity and abundances in Norwegian fjords, particularly amongst species with low mobility. The full impact on the ecology of Norwegian coastal waters is still unknown, but a loss in diversity, benthic production and changes in nutrient recycling will probably have negative impacts on fish populations.

FIRE CRAB

Pyromaia tuberculata (family Inachoididae) is a small crab originally native to the Pacific American coast from San Francisco to the Panama Canal. Not content to stay at home, it is now found in Japan, Brazil, Argentina, southern Australia and New Zealand, probably transported as larvae in ballast water. This species is very efficient at reproduction – females can carry fertilized eggs within days of their puberty moult, they can reproduce all year round, and larval development is faster than that of many other crabs. This no doubt gives them a competitive advantage over similar native species in their new environments. The fire crab lives on temperate rocky reefs amongst seaweeds, sponges, and other fouling, and is fond of wharf piles. An ecological study in Port Phillip Bay, Australia, indicates that it has become an important part of the diet of local bottom-feeding fishes.

SNOW CRAB

This large 'spider crab' (*Chionoecetes opilio*; family Oregoniidae) occurs naturally in Arctic waters of the northern Pacific, the Beaufort Sea, western North Atlantic, and the west coast of Greenland. It was first recorded as an invasive species in the Barents Sea by Russian vessels in 1996, where it was presumed to have been a ballast-water introduction. Numbers have continued to increase, and since about 2013 it has become an important new commercial stock, particularly for the Norwegians. The new Barents Sea stock has even become a source of bitter dispute between Norway and the European Union, with the EU believing it should be freely available for capture by its fishers within international waters, while Norway argues that the crab falls under its jurisdiction over its entire continental shelf.

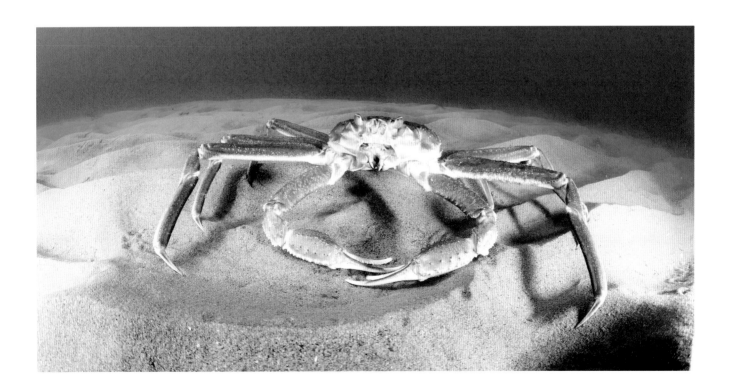

CRAB CONSERVATION

WITH OUR RAPIDLY CHANGING WORLD, crabs, along with a host of other animals and plants, are finding themselves in a fight for their very existence.

CLIMATE CHANGE

Increasing levels of carbon dioxide from burning fossil fuels is steadily, and measurably, changing the fundamental chemistry of our oceans. Too much CO_2 leads to the formation of carbonic acid in the water and reduces the availability of carbonate, a critical component of shell-building. This will increasingly impact on invertebrates that need calcium carbonate for their shells, such as crustaceans, corals, molluscs and echinoderms. Their tiny, delicate planktonic larval stages are especially vulnerable. If larvae do not survive to adulthood, this will not only affect the recruitment of new generations, but because plankton communities are at the base of marine food webs, the whole oceanic ecosystem will be disrupted, and fish stocks will become depleted. A new prescient study has found that some larvae of Dungeness crabs in the US Pacific Northwest are already smaller than usual, and are suffering pitted and folded shells, as well as damage to tiny sensory setae that could lead to such issues as slower movements and impaired swimming.

Warming of the oceans is an equally serious threat. Corals are bleaching and dying across huge areas, and without adequate and timely recovery, there will be a devastating cascade effect on the myriad invertebrates that depend on this biological system. Warmer seas will also have the effect of speeding up metabolism and therefore the need for oxygen, but warmer water also has less ability to hold oxygen, so larger marine animals may begin to move away from equatorial waters due to respiratory stress. A temperature rise of several degrees (as predicted by the end of this century) will mean sea water holds 5 to 10 per cent less oxygen than it does currently.

Models of the impact of this on populations of Atlantic rock crab (*Cancer irroratus*) indicate that viable habitats will be restricted to shallower, more oxygenated surface water, and around 20 per cent of the population will be forced away from an uninhabitable equatorial region. Warming also means melting polar ice caps and rising sea levels. Sea levels have risen before, but not at the rapid rates being forecast, and it is likely that many coastal ecosystems will not have time to adapt. Intertidal mangrove swamps and their rich diversity of crabs and other specialized fauna and flora will be submerged, as will beaches that have formed over millennia.

THE FUTURE FOR FRESHWATER CRABS

Many crabs, especially marine species, have broad distributions, and are able to spread their larvae far and wide on the currents – such species have a natural resilience to local threats. However, others are much more localized and ecologically specialized. This applies very much to the multitude of freshwater crabs that lack larval dispersal (baby crabs are directly hatched). Many of these species are confined and endemic to single river catchments. Freshwater crabs are a key component in many tropical freshwater systems. They are herbivorous or omnivorous scavengers and predators. They are also an important source of food for a range of birds, reptiles and mammals. However, rapid habitat deterioration through deforestation, soil erosion, increasing stream-silt loads, alteration of drainage patterns, pesticides and pollution, means that many species are now under imminent threat. In 2009, an IUCN Red List assessment of the 1,280 species then known worldwide, noted that two-thirds of all species may be at risk of extinction, with one in six species particularly vulnerable. Of those, 227 species should be considered as Near Threatened, Vulnerable, Endangered or Critically Endangered. The proportion of freshwater crabs threatened with extinction exceeds that of all other groups assessed except for amphibians.

In Malaysia, and in most parts of Southeast Asia, the widespread loss of natural forest, as a result of land development and agriculture, has impacted almost every habitat where freshwater crabs are found. In Sri Lanka, freshwater crab species are restricted to montane and sub-montane habitats where there has been so much ecological disturbance that less than 20 per cent of all the island's species are still considered to have relatively secure populations.

Climate change is also likely to have an important impact on freshwater crabs in Australia. New research indicates that most small coastal flowing catchments across northern Australia have their own endemic species. In an earlier era, when the region was hot and wet, and rainforests flourished, the crabs spread everywhere, but as the climate became arid for much of the year, the crabs were restricted to their individual river systems. If under some climate change scenarios, prolonged droughts were to last at least several years, it seems probable that many species that have evolved under an annual monsoonal regime, will not have the resilience to survive. Like most freshwater crabs across the world, almost nothing is known of their biology, ecology, behaviour or physiology.

LEFT: When photographed on Christmas Island in 2013, this white-stripe crab (*Labuanium vitatum*) had not been seen there for 25 years. It must be considered highly vulnerable to extinction.

RIGHT: *Yuebeipotamon calciatile* is found only in the limestone hills around Yingde City in China – typical of the limited distributions of many freshwater crabs.

CHRISTMAS ISLAND: WHERE CRABS RULE!

This far-flung Australian Territory in the northeastern Indian Ocean is a tiny patch of extraordinary life. Eroded limestone reef terraces encircle the basalt tip of an ancient volcanic seamount that emerges sharply from the abyssal ocean plain 4.5 km (almost 3 miles) below. It has been isolated for so long that a rich and unique fauna has evolved. Indeed, it has been called the 'Galápagos of the Indian Ocean' because of its large number of indigenous species. But what makes Christmas Island so special is the fantastic variety (over 20 species), and the sheer numbers, of terrestrial or semi-terrestrial crabs – easily more than anywhere else in the world. Even more uniquely, crabs are the keystone organisms that govern the forest structure and ecology of the whole island.

The most famous of the indigenous species is the Christmas Island red crab (*Gecarcoidea natalis*), whose massed annual migration is one of the great natural events. Millions of crabs, triggered by the start of the summer rains, migrate to the sea in a seething, red carpet of life. Their mission is simple: to cast their eggs into the water and ensure new generations of Christmas crabs. But while the existence of the red crab is still considered safe, it cannot be taken for granted. Its greatest immediate threat (and also to other crabs on the island) is the yellow crazy ant (*Anoplolepis gracilipes*), one of the 100 worst invasive species on Earth. These extremely aggressive ants were accidentally introduced to Christmas Island around 75 years ago. Despite having no natural

ABOVE RIGHT: A seething carpet of Christmas Island red crabs form a blur of colour as they march relentlessly across the landscape, driven by the fundamental need to reproduce.

RIGHT: Crab bridges on Christmas Island were designed to protect red crabs from being killed by traffic. Previously it is believed that up to a million crabs were killed every year, especially during migration.

200   EXPLOITATION AND CONSERVATION

enemies on the island, they curiously persisted in only relatively low numbers. However, in 1989 the first 'supercolony' was found, and the sleeping giant was awoken with a terrible resolve.

The supercolonies on Christmas Island have multiple queens, as many as 11 nests per square metre, and can cover areas up to 750 hectares (1,800 acres) at a density of 20 million ants per hectare – the highest density of foraging ants ever recorded. They kill crabs using formic acid, which is squirted from the tip of the abdomen, often initially blinding them before they are overwhelmed by sheer numbers! Since their major population explosion in the mid-1990s, yellow crazy ants have been responsible for the demise of at least 10 to 15 million adult red crabs, and countless numbers more of the highly vulnerable juvenile crabs returning from the sea (as few as one or two ants can kill a baby crab). Baiting programmes have been helpful but not sustainable in the long term, so an attempt at bio-control is currently being undertaken using a tiny Malaysian wasp (*Tachardiaephagus somervillei*), only about 2 mm ($1/12$ in) long. The ants need a steady supply of honeydew that is mostly produced, ironically, by another introduced species, the yellow lac scale insect. The wasp preys solely on this scale insect, piercing its body to lay its eggs, so that it becomes a meal for the newly hatched wasp larvae. Without a good supply of honeydew, it is hoped that the crazy ants will be brought back under control – results so far are very promising.

Red crabs (and the island's robber crabs) also face serious human-induced problems. Cars and mining trucks crush them by their hundreds of thousands as they cross the roads, especially during their mass migrations. As early as 1983 it was estimated that as many as one million crabs were being killed annually by the heavy traffic, and by 1996 around 16 per cent of all crabs taking part in the migration were dying. Following the construction of the Immigration Detention Centre (2002–18), mortality must have increased. Ongoing work to largely mitigate this carnage has involved road closures during the peak migration periods, and low fences designed to direct crabs to safe passages below the road surface, or even onto specially designed bridges.

Climate change may be the biggest threat of all. Increasingly unreliable wet seasons can see the crab migration fail, or the crabs start on their migration only to die from dehydration away from the protection of their burrows, if rain fails to continue. In the absolute worst scenario, if there were ever changes in oceanic circulation patterns around this remote island, the crab larvae of all the indigenous species could be swept away, never to return.

LEFT: The stunning sky-blue *Discoplax celeste* (family Gecarcinidae) is one of a number of semi-terrestrial crabs indigenous to Christmas Island.

RED KING CRAB
Paralithodes camtschaticus

FAMILY:	Lithodidae
OTHER NAMES:	Alaskan king crab, Kamchatka crab, Stalin's crab
DISTRIBUTION:	North Pacific: Alaska south to Queen Charlotte Islands; south to Japan. Introduced to Barents Sea; spreading south in north Europe.
HABITAT:	Sand or mud bottoms to 370 m (1,200 ft), but mostly less than 200 m (660 ft); juveniles in intertidal, shallow subtidal rocks
FEEDING HABITS:	Generalist predator: eating anything from small invertebrates to large echinoderms and bivalves
NOTES:	Females stay in warmer water (4°C/39°F) to promote egg development; males prefer colder (1.5°C/35°F) to conserve energy.
SIZE:	To about 29 cm (11½ in) carapace width

EATING MACHINES

Large numbers of red crabs can eat vast quantities of bottom-dwelling invertebrates. In the waters of northeastern Europe, where there are no natural predators, burgeoning numbers of these crabs threaten the ecological balance of coastal shelf waters.

THE RED KING CRAB HAS, FOR MANY YEARS, been the basis of a lucrative fishery in the coastal waters of its native north Pacific, around the Kamchatka Peninsula and neighbouring Alaskan waters. It is the largest species in the fishery, reaching a leg span of 1.8 m (6 ft) and weighing more than 10 kg (22 lb). It is highly valued as food and exported throughout the world; the Russian fishery alone is worth US$200 to 250 million per year. Populations continue to decline in Alaskan waters, probably because of overfishing and warming sea temperatures, and tighter fisheries controls have so far failed to arrest the decline.

THE COLD WAR CONTINUES!

To increase productivity, Soviet scientists introduced the red king crab into the Atlantic in 1960, initially along the coast of northwestern Russia (hence the name 'Stalin's crab'). It was found in Norway in 1977, and by the 1990s it had become well established. Most fjords in northern Norway are now home to the occupying forces. Worse still, they appear to be continuing to move south, with warmer sea temperatures seemingly no barrier. It is estimated that there are now over 20 million crabs in the Barents Sea, and divers regularly see pods of 10,000 young crabs swarming over the bottom. Spawning, moulting and mating occur in spring, when crabs migrate into shallow waters.

SPANNER CRAB
Ranina ranina

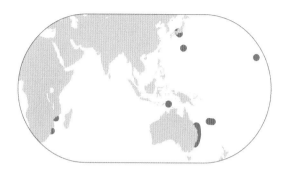

THE PECULIAR-LOOKING SPANNER CRAB takes its common name from the shape of its claws – flat fingers turned inwards at 90 degrees resemble the jaws of a wrench. It prefers to live on open bottoms in sandy offshore areas, burying itself with just its eyes and the top of its mouth protruding. Here it waits to attack passing prey – small, bottom-dwelling fishes and invertebrates. Interestingly, during breeding season, spanner crabs have been reported to swarm in large numbers onto exposed sandbanks and beaches at low tide.

SUSTAINABLE PRACTICES

The largest fishery is along the mid-eastern coast of Australia. Crabs are harvested year-round, barring the summer spawning season (20 November to 20 December). Commercial crabbers are only allowed to use dillies (frames with netting stretched across them) no bigger than one square metre. A bag of bait fish is attached to each dilly, 10 or 15 dillies are clipped together, and the gear is left in the water up to an hour, then winched aboard. Berried females and undersized crabs (under 10 cm [4 in] in carapace length) are returned to the water, while the remainder are kept under shade and frequently sprayed with seawater. Most of the catch is exported live to East Asia (mainly Taiwan), although a small quantity is sold locally as chilled cooked product.

FAMILY:	Raninidae
OTHER NAMES:	Red frog crab
DISTRIBUTION:	Widespread across the Indo-West Pacific region
HABITAT:	Intertidal zone to more than a 100 m (330 ft) depth; prefers bare sandy areas where it can easily bury itself
FEEDING HABITS:	Ambush predator and scavenger
NOTES:	*Ranina* is a derivation of the Latin *rana*, meaning 'frog', alluding to the crab's supposed frog-like appearance
SIZE:	To about 15 cm (6 in) carapace width, and about 900 g (2 lb) in weight; females are smaller, at about 11.5 cm (4½ in) width

OLD IS GOOD

Spanner crabs belong to the Archaeobrachyura, an ancient lineage whose origins date back to the beginnings of crab evolution. Perhaps this helps explain its strange appearance. It is also the only representative of these primal early crabs to be commercially harvested.

TIGER CRAB
Orithyia sinica

FAMILY:	Orithyiidae
OTHER NAMES:	Tiger-face crab
DISTRIBUTION:	Coast of mainland Asia, from South Korea to Hong Kong
HABITAT:	Rocky sandy to sandy-mud bottoms, in shallow inshore waters
FEEDING HABITS:	Predator and scavenger
NOTES:	In Greek mythology, Orithyia was one of the daughters of Erechtheus, King of Athens
SIZE:	To 65–70 mm (2½–2¾ in) carapace length

THE TIGER CRAB GETS ITS NAME from its reddish-brown striped legs, and the large 'eye' spots that stare out from either side of its carapace. Genetic analyses reveal that *Orithyia sinica* represents an old lineage and has remained much the same for millions of years, confirming older studies that have failed to find close relatives amongst other crabs. Little is yet known about its biology and ecology. In aquarium studies, crabs prefer cleaner sandy substrates over muddier ones, and they can quickly bury backwards using the spatulate claws on their last two pairs of legs. Even so, they never dig deep enough to fully cover themselves, and prefer the protection of rocks and rubble. They are predators and probably scavengers, likely preferring small molluscs, worms and other crustaceans.

RESTRICTED HABITATS

Curiously, the tiger crab is confined to the continental margin and completely absent from nearby islands, such as Taiwan, the Ryukyu Islands and Japan. Each female produces from 25,000 to 100,00 eggs, and the larvae are planktonic; however, there are only three larval stages, so perhaps the period of larval development is foreshortened, and local current patterns keep the larvae close inshore. There is also evidence that adults prefer periods of lowered salinity that would occur from mainland run-off.

TIGER'S EYES FOR CAMOUFLAGE

This species is the basis of a small but valuable fishery using tangle nets. The crabs appear to like half-burying in sand amongst rocks and rubble. The pair of large brown 'tiger-eye' spots on rounded lumps must appear to predators as simply the tips of small protruding rocks.

COMMON EUROPEAN SPIDER CRAB
Maja brachydactyla

ONE OF THE MAINSTAYS OF THE EUROPEAN seafood diet since hominids first arrived from Africa, the identity of this crab has nevertheless been confused until recently. Long known as *Maja squinado*, it is now realized that two similar species had been mixed together. *Maja squinado* is confined to the Mediterranean, and although once abundant, is now endangered. *Maja brachydactyla*, conversely, has a wide distribution along the East Atlantic coast, from the British Isles to West Africa.

A COMPLICATED LIFE CYCLE

Freshly settled juvenile crabs live in the shallows for about two years before reaching sexual maturity. Once adult, they venture out into deeper waters. Mature crabs then undertake a summer migration back into the shallow waters, often forming mounds or pods ranging in size from a few dozen, up to 50,000 individuals! These masses of moving crabs appear to serve as a way to protect each other – the crabs on the surface of the pods still have hard shells, whereas those in the centre are soft from just beginning or finishing their moulting and so are vulnerable to a hungry predator. But it is also a time when males mate with the females while they are still soft from casting off their old shells.

FAMILY:	Majidae
OTHER NAMES:	Common spider crab
DISTRIBUTION:	Northeast Atlantic south to Namibia
HABITAT:	Muddy-sand to rock bottoms; subtidal to more than 100 m (330 ft) depth
FEEDING HABITS:	Omnivorous, but seaweeds (macro-algae) form a major component of the diet, supplemented by a variety of small invertebrates
NOTES:	*Brachydactyla* is Latin for 'short fingers'
SIZE:	To about 22 cm (8¾ in) carapace length

A QUESTION OF IDENTITY

Genetic techniques allow us to more easily identify cryptic species, and this can have important consequences. *Maja squinado* of the Mediterranean, with which this species has been long confused, is now endangered and needs separate management practices from its widespread doppelganger.

HORSEHAIR CRAB
Erimacrus isenbeckii

FAMILY:	Cheiragonidae
OTHER NAMES:	Hair crab, queen crab
DISTRIBUTION:	Northern Pacific: Korea, Japan and Kamchatka Peninsula to Aleutian Islands and Gulf of Alaska
HABITAT:	Open sand to silt bottoms at 30–350 m (100–1,150 ft) depth
FEEDING HABITS:	Eats other crustaceans, including isopods, and other small invertebrates
NOTES:	Known to cannibalize smaller members of its own species; preyed on by flatfish and salmon
SIZE:	To about 12 cm (4¾ in) carapace width

CONFINED TO COLD TEMPERATE and boreal waters (2–4°C [36–39°F]) of the northern Pacific, the horsehair crab is the single representative of its genus, and one of only three species in its family. Attaining a kilogram in weight (2¼ lb), it is particularly prized in Japanese cuisine.

SLOW GROWERS

There have been comparatively few studies of its biology, but it is known that they take four years to reach sexual maturity. This slow growth rate seems primarily controlled by low water temperature. They also reproduce only every three years, with each female typically carrying 40,000 to 50,000 eggs, although the largest crabs can have as many as 160,000. Embryonic development is slow, and it can be a year between mating and the settlement of baby crabs. Hatching and release of the planktonic larvae occurs from March to May to coincide with the northern hemisphere's spring phytoplankton blooms. Settlement to the bottom as megalopae finally occurs around July.

Like other slow-growing and long-lived species, horsehair crabs are vulnerable to over-exploitation. Since a peak of 25,000 tonnes per annum in the 1950s, less than 1,800 tonnes are taken today; it is now well managed, with strict catch-limits in place and some stock-enhancement programmes initiated. However illegal harvesting in Russian waters is of significant concern.

DANGERS OF OVERFISHING

It is vitally important to understand the reproductive biology and ecology of a species if it is to be sustainably fished. Horsehair crabs are slow-growing, long-lived, and each female only spawns every three years, with a relatively small number of eggs.

CHINESE MITTEN CRAB
Eriocheir sinensis

The Chinese mitten crab plays both hero and villain. Native to the northwestern and western Pacific (Southeast Asia, China and Korea), it is much relished as a delicacy in that region, and the Chinese aquaculture industry produces more than 700,000 tonnes a year. However, this species could also be the poster child for all would-be crab invaders, truly setting the bar high. It is regarded as one of the top 100 invasive species by the International Union for Conservation of Nature and Natural Resources.

RETURNING TO THE SEA

Mitten crabs have a catadromous life cycle, meaning that they spend most of their lives in fresh water, but sexually mature crabs must migrate to the sea every summer to reproduce. They never make the return trip, with both sexes dying after the eggs have been released. Females can produce from 250,000 to 1 million eggs, depending on their size and condition. Baby crabs finally make the arduous journey back through the estuary to fresh water, where they will live and grow for the next one to five years, before themselves returning to the sea.

FAMILY:	Varunidae
OTHER NAMES:	Asian mitten crab, Chinese freshwater edible crab, Chinese river crab
DISTRIBUTION:	Originally native to Southeast Asia, China and Korea; now circum-global across the northern hemisphere
HABITAT:	Subtropical to temperate freshwaters; riverine
FEEDING HABITS:	Omnivorous: consumes detritus, aquatic plants and other small invertebrates
NOTES:	A secondary intermediate host of the Oriental lung fluke (*Paragonimus westermani*), a serious health issue for humans
SIZE:	To about 80 mm (3 in) carapace width

THE MOST SUCCESSFUL CRAB?

Mitten crabs are the most farmed crabs; one of the most invasive, physiologically tolerant and ecologically damaging of any species; a major vector of disease; hugely prolific breeders, with juveniles able to migrate up to 1,500 km (930 miles); and they have such beautiful hairy claws!

EUROPEAN EDIBLE CRAB
Cancer pagurus

FAMILY:	Cancridae
OTHER NAMES:	Brown crab, Cromer crab, *tourteau* (French), *buey de mar* (Spanish)
DISTRIBUTION:	Northeast Atlantic coasts, from Norway south to Morocco; northern Mediterranean Sea
HABITAT:	Found from the lower intertidal zone to 100 m (330 ft) depth. Lives under rocks or in cracks and holes, but also buries in course to muddy sand substrates during the day.
FEEDING HABITS:	Nocturnal predator on molluscs and other crustaceans, such as crabs and squat lobsters; also an active scavenger
NOTES:	Typically lives for 25–30 years, but may reach 100 years
SIZE:	To 30 cm (12 in) carapace width, and up to 3 kg (6 2/3 lb) in weight

THE EUROPEAN EDIBLE CRAB has no doubt been a prized menu item since our human ancestors first migrated to coastal Europe from Africa. It was also one of the first crabs to be given a modern scientific name by Carl Linnaeus in 1758. *Cancer* simply means 'crab', and *pagurus* is a variation on the ancient Greek *pagouros* (from *pagos* for 'rock', and *uros*, meaning 'to fix or fasten'), so its binomial epithet simply translates as 'crab associated with rocks'.

MILLIONS OF EGGS

This crab is the most commercially exploited crab species in Western Europe, with as much as 60,000 tonnes caught annually. There is some concern that *Cancer pagurus* is being overfished, at least in the waters around the United Kingdom and Ireland, but it is still a largely viable fishery. This resilience despite such large catches is no doubt because individual crabs can live to 100 years old, and each female produces between a quarter of a million and 3 million eggs a year, depending on her size. Crabs are caught using offshore baited pots, but also in trawl nets.

THE ICONIC CRAB OF EUROPE

This large coastal crab is commonly found living under rocks and in crevices in the lower intertidal zone. A popular food item, females are often referred to as 'hens', while males are 'cocks'.

LITTLE RED VAMPIRE CRAB
Geosesarma hagen

GEOSESARMA IS A GENUS OF SMALL semi-terrestrial to terrestrial crabs that evolved from relatives that inhabit coastal mangroves. They live through the tropical Indo-Malayan region, and at least 60 species have now been described, with more known to exist. Most have narrow distributional ranges and specific ecological habitat requirements, making them vulnerable to human encroachment and exploitation. Only formally named in 2015, the beautiful little red vampire crab is restricted to a very small area in central Java, where there is lots of trickling water and small rivulets. Adults hide amongst thick ground vegetation and rocks, and in burrows along the edge of streams, although they seldom enter the water itself. Juveniles are more aquatic, staying close to the waterline, where they can make a quick underwater escape from terrestrial predators.

AQUARIUM FAVOURITES

Geosearma hagen and many related species have bright, distinctive colours that make them popular with aquarium enthusiasts, so they are collected in large numbers for export. As a result, populations are decreasing. Their conservation needs have not yet been assessed, but given their limited distributions, many will soon be under threat. Sustainable harvesting and captive breeding programmes to supply the demands of the trade will become increasingly important.

FAMILY:	Sesarmidae
OTHER NAMES:	Red devil crab
DISTRIBUTION:	Central Java, Indonesia
HABITAT:	Terrestrial as adults; hide under thick ground vegetation and rocks, sometimes excavating burrows
FEEDING HABITS:	Carnivore, scavenger, insectivore
NOTES:	The extent of bright reddish orange on the carapace varies, with some specimens being almost completely orange
SIZE:	To about 14 mm (1/2 in) carapace width

FOREST COLOURS

Many vampire crab species are uniquely and spectacularly coloured. While they seem likely to draw a predator's attention, they are without doubt reflecting the colours of the rainforest bark, leaves, fruits and flowers that surround them.

GLOSSARY

BATHYAL An ocean depth zone typically extending from the edge of the continental shelf at about 200 m down to 2,000 m (660–6,600 ft) below the surface.

BENTHIC Animals that live on the bottom under water; the opposite of pelagic, where the animal lives by swimming or floating near the surface.

BERRIED Refers to female crabs that are carrying a large clutch of eggs under their pleon; the appearance is reminiscent of a bunch of berries growing on a vine.

CARCINIZATION The evolutionary process of becoming a crab or 'crab-like'. It involves the pleon (abdomen) folding tightly beneath the body with its primary function reduced to reproductive processes; a hardening of the shell; and a general widening of the body with many internal changes.

CHROMATOPHORES Pigment structures that help give crabs their distinctive colours. They are mostly located in the epidermis, and classified by colour: melanophores (black/brown), leucophores (white), erythrophores (red) and xanthophores (yellow).

COMMENSAL Animals living in symbiosis where neither seems to obviously gain or lose anything because of the relationship.

DECAPOD Crustaceans belonging to the Order Decapoda, each having ten legs arranged as five pairs (including the claws). Decapods include most of the common larger crustaceans encountered, such as crabs, prawns, shrimps and lobsters.

DIRECT DEVELOPMENT In most crustaceans, eggs hatch into free-living larval forms that undergo a series of moults before turning into a juvenile form of the adult. With direct development, larval stages progress while still encased within the egg, or under the female abdomen, before hatching as a megalopa or juvenile.

ECOLOGICAL CONVERGENCE Occurs when two unrelated species look very similar in appearance because both have evolved independantly to suit similar ecological niches.

INTERTIDAL The strip of shoreline that is exposed daily by the ebb and flow of the tides. It is defined as the zone between the lowest neap tide and the highest spring tide each month.

MASS MOULTING The phenomenon of crabs congregating in large numbers before undergoing moulting together. Believed to provide safety in numbers.

MEGALOPA The last larval stage of a developing crab, and the first stage to settle out of the plankton onto the bottom. It moults into the first juvenile crab stage.

MUTUALISM When two different types of animal live together in a relationship where both benefit by the association.

NAKED RETINA A unique visual adaptation occurring in hydrothermal vent crabs (Family Bythograeidae) that allows them to detect near-infrared portions of the light spectrum produced by vent chemistry. Light enters through a transparent, non-faceted cornea.

NEUSTONIC Tiny organisms that inhabit the zone on or just below the surface of the water.

OBLIGATE SYMBIONT An animal that is dependent on the services that its symbiotic partner provides, and cannot live freely on its own for any length of time.

SETAE Hair-like extensions of the cuticle that are characteristic of crustaceans. They can vary from 10 to 20 microns to over 10 mm in length; be fragile and slender to strong and robust; and are used for everything from camouflage to feeding, grooming and sensing the environment.

SEXUAL DIMORPHISM Differences between the sexes within a species, particularly in relation to secondary sexual characteristics at maturity. In a crab it could refer to consistent differences in claw size, or in colour, amongst other things.

SPECIATION The evolutionary process of becoming separate species. Most commonly occurs when populations become isolated and can no longer interbreed (allopatry).

SUBTIDAL Any marine habitat that is below the lowest tidal limit. Most often used to refer to relatively shallow waters, but no defined limit.

SUSPENSION FEEDER An animal that feeds by sieving or straining food particles that are suspended in the water column around it. Relatively rare amongst brachyurans.

SYMBIONT/SYMBIOSIS Symbiosis simply means the 'living together of dissimilar organisms'; it includes a variety of types of relationships, including parasitism, commensalism and mutualism.

FURTHER READING

BLISS, D. E. (1982). *Shrimps, Lobsters and Crabs: Their Fascinating Life Story*. Piscataway, NJ: New Century Publishers.

BURGGREN, W. W. and MCMAHON, B. R., eds (1988). *Biology of Land Crabs*. Cambridge, UK: Cambridge University Press.

CASTRO, P., DAVIE, P. J. F., GUINOT, D., SCHRAM, F. R. and VON VAUPEL KLEIN, J. C., eds (2015). 'Decapoda: Brachyura (Parts 1 & 2). Treatise on Zoology – Anatomy, Taxonomy, Biology.' *The Crustacea*, vol. 9C-I, 638 pp., vol. 9C-II, 584 pp. Leiden and Boston: Brill.

JENSEN, G. C. (2014). *Crabs and Shrimps of the Pacific Coast: A Guide to Shallow-water Decapods from Southeastern Alaska to the Mexican Border*. Bremerton, WA: Mola Marine.

ORCHARD, MAX (2012). *Crabs of Christmas Island*. Christmas Island Natural History Society.

POUPIN, J. AND JUNCKER, M. (2010). *A Guide to the Decapod Crustaceans of the South Pacific*. Noumea, New Caledonia: CRISP and SPC. (Downloadable at www.crisponline.net)

WARNER, G. F. (1977). *The Biology of Crabs*. London: Elek Science.

WEISS, JUDITH S. (2012). *Walking Sideways: The Remarkable World of Crabs*. Ithaca, NY: Cornell University Press.

INDEX

Numbers in *italic* refer to pictures or their captions.

A
Acanthodromia 50
  A. erinacea 50
  A. margarita 50, *51*
Achaeus japonicus 10
acron 64
Adam's zebra crab *140*, 141
Aethra scruposa 159, *159*
Aethridae 18, 67
Aethroidea 18
Afruca tangeri 83
Albunea symmysta 27
Albuneidae 26
allopatric speciation 39
ambulatory legs *63*, 65
ambush predators 6, 58, *104*, 105–6, *105*, 158, 204
amphibious crabs 80, *80*, 118
anemone crab *see* boxer crab; *Lissocarcinus laevis*
Anomura 14, 26–9, 32, 35, 46, 49
antennae 22, *63*, 64, *64*, 81
antennules *63*, 64, *64*, 106
antlered crab *33*
Aphanodactylidae 19, 120
arachnids 22, 25
Arcania
  A. septemspinosa 67
  A. undecimspinosa 75
Archaeobrachyura 14, 32, 34, 35, 53, 58, 72, 94, 144, 161
Aristotle 10–11
arrow crab *84*, 85
arrowhead crab *see Huenia heraldica*
Arthropoda 22
Ashtoret lunaris 67
asymmetry 90, *91*
Atelecyclidae 18
Atergatis
  A. floridus 192, 193
  A. roseus 194
Attenborough, Sir David 16
Austinograea hourdezi 128, 129
Australocarcinus 153
  A. riparius 119
Austrothelphusa 42, *42*
  A. transversa 80, *80*, 118, 154
Austruca
  A. bengali 83
  A. lactea 83
autotomy 70

B
bad-hair-day crab 54, *55*
Baird's tanner crab *see Chionoecetes bairdi*
balance, sense of 64, *64*
Barth's organ 157
Baruna socialis 67
basis-ischium 65, 70
batwing crab 115
Beebe, Charles William 8
Belliidae 18
Bellioidea 18
bioturbators 105
Birgus latro 28, *44*, 45, 182, *182*, 201
black-fingered crab 93
blackfly, biting 191, *191*
bladder 64, 77
blue crab
  Callinectes sapidus 106, *109*, 183
  common 106
  Japanese 183
  large *109*
  Malawi 118
blue swimmer crab *see Portunus armatus*
box crab *see Calappa*
boxer crab *see Lybia*
brachiopods 30
Brachyura 12, 30, 32
brain 81, 156–7
branchial *63*
bromeliad crab *see Metopaulias depressus*
brown crab *see Cancer pagurus*
Brusiniidae 19
buccal cavity *63*
Buddhist crab 172, *173*
burrowing species 26–7, *43*, 53, 64, 65, 67, 69, *69*, 89, 101, 105, 145, 149 182
Bythograeidae 18, 74, 117, 129
Bythograeoidea 18

C
Cacinidae 195
Calappa 68, 104, 121, 163
  C. calappa 66
  C. gallus 82
  C. lophos 162, 163
  C. ocellalta 111
  C. philargius 67
Calappidae 18, *67*, 163, 164
Calappoidea 18
calcium 31, 75, 80, 154
calcium carbonate 62, 198
Callichimaera perplexa 31
Callinectes sapidus 106, *109*, 183

Calvactaea tumida 149
Cambrian 22, 30
camouflage 54, 58, 69, 82, 116, 158–9, 207
  Calappa lophos 162, 163
  carrying behaviour 12, *13*, 28, 66, 71, 94, 95, *115*, 134, *135*, 159–61, *160*, *161*, 168, *169*
  crypsis 159–60
  Hoplophrys oatesii 166, 167
  mimicry *159*, 160, *160*, *166*, 167, *170*, 171
  sculptured crab 138
Camposcia retusa 168, *169*
Camptandriidae 19, 67, 73
Cancer 11
  C. irroratus 76, 198
  C. magister 183, 192, *193*, 198
  C. pagurus 64, 70, 77, 183, 193, 214, *215*
  C. productus 193
Cancer constellation 8–9, *8*
Cancridae 18, 215
Cancroidea 18
candy crab *166*, 167
carapace 12, 62, *62*, *63*, 77
Carcinidae 19
carcinization 26, 29, 81
Carcinoplax purpurea 73
Carcinus maenas 76, 78, 81, 104, 149, 156, 193, 195, *195*
Cardisoma guanhumi 181
Carpiliidae 18
Carpilioidea 18
Carpilius corallinus 115
carpus *63*, 66
carrying behaviour 12, *13*, 28, 66, 71, 94, 95, *115*, 134, *135*, 159–61, *160*, *161*, 168, *169*
catadromous species 212
cave-dwelling species 74, 119, *119*, *124*, 125
Celsus, Aulus Cornelius 8
cephalon 62
cephalothorax 62, 69, 78
Ceratocarcinus longimanus 120
Chaceon
  C. albus 184
  C. fenneri 184, *184*
  C. notialis 184
  C. quinquedens 184, *187*, 188
  C. ramosae 188
champagne crab 188
Charybdis natator 144
Chasmocarcinidae 153
Cheiragonidae 18, 144, 211
Cheiragonoidea 18

chelae 12, *63*, 66–8, *66*
Chelicerata 22
chelipeds 22, *63*, 65, 66–8, *66*, *67*, 70, 76, 97
Chionoecetes
  C. bairdi 147, *186*, 192
  C. opilio 180, 183, 192, 197, *197*
Chlorodiella nigra 115
Christmaplacidae 19, 125
Christmaplax mirabilis 124, 125
Christmas Island *199*, 200–1
Christmas Island blind cave crab *124*, 125
Christmas Island red crab *see Gecarcoidea natalis*
chromatophores 82
Ciliopagurus tricolor 46, *47*
circulatory system 78, *79*
classification 11, 18–19, 34–5
claws *see* chelipeds
climate change 182, 198, 199, 201
Clistocoeloma 159
coconut crab *see Birgus latro*
Coenobita perlatus 45
Coenobitidae 28, 45
cognition 81, *81*, 156–7
colours and patterns 81, 82, *82*, *83*, 89, 93, *93*, 98
commensal species *12*, 85, 89, 115, 120, *120*, *123*, 141, *149*, *166*, 167
communication 82, 101, 156–7, *157*
conical gorgonia crab *159*
conservation 198–201, 204
copepod crustaceans 22
coral 7, 30, 31, 32, *36*, 97, 104, 115, 198
  commensal species 89, 115, *141*, *149*
  mucus 50, 57, 68, 104, 120, 167
  symbiotic species 57, 115, *115*, 120, *120*, *159*, 160
Corycodus 58
  C. disjunctipes 58, *59*, 67
Corystidae 18, 64
Corystoidea 18
coxa *63*, 72
Crab Nebula 10, *11*
Cranuca inversa 83
crayfish, red swamp 118
Cretaceous 15, 31, 33
Cretaceous–Paleocene extinction event 32
Cromer crab *see Cancer pagurus*
Crossotonidae 160

Crossotonotus spinipes 160
Crustacea 10, 12, 22
crustacyanin 82, 89
crypsis 159–60
Cryptochiridae 12, 19, 57, 120, 121
Cryptochiroidea 19
Cryptopodia dorsalis 42
crystal crab 184
cuticle 62, 65, 75, *75*
Cyclocoeloma tuberculata 13
Cyclodius ungulatus 115
Cyclodorippidae 18, 58, 67
Cyclodorippoidea 18
Cymo 120
Cymonomidae 18
Cyrtomaia 116, *116*

D
dactylus *63*, 66
Dagnaudus petterdi 33
Dairidae 18
Dairoidea 18
Dairoididae 19
Danarma garfunkel 176, *177*
Dardanus pedunculatus 28
Decapoda 22, 65
decorator crab *13*, 151, 160, *160*, 168, *169*
deep-sea dwellers 15, 32, 74, *74*, 107, *107*, 112, 116–17, *116*, *117*, *128*, *129*, 184
defence 22
  boxer crabs *15*, 66, 86, *87*, 121
  carrier crabs 161, 168
  spines 96
  toxic species 89, *92*, 93, 192–3
Demania splendida 193
Dendronephthya 166, 167
desert conditions 15, 176
detritivores 7, 32, 68, 104, 118
devil crab 193
Dicranodromia chenae 71
diet 22, 66, 68, 76, 104
  algae 68, 76, 93, 97, 104, 108, 110, 114, *114*, 115
  coral mucus 50, 57, 68, 104, 120
  detritivores 7, 32, 68, 104, 118
  dietary pigments 82, *93*, 98
  filter feeders 26, 30, 104
  parasitic species 141
  predatory species *6*, 7, 58, 66, 68, 90, 104–5, *104*, *105*, 106, *107*, 118, 158, 204
  scavengers 30, 32, 45, 104, *107*, *107*, 193

suspension feeders 27
symbiotic species 120–3
vampire crabs 118–19, *119*, 216
vegetarian species 108
digestive system 76, *77*
Diogenidae 46
Discoplax celeste 201
disease, food-borne zoonosis 190–1, *190*, *191*
dispersal 38–9, *38*, *39*, 42
display behaviour 145, *145*, *172*, *173*, *175*
Domeciidae 19, 120
dominance hierarchies 158, *158*
domoic acid 192, *193*
Dorippe quadridens 161, *94*, *95*
Dorippidae 65, 94
Dorippoides facchino 161
Dotilla 181
Dotillidae 19, *67*, *80*, *172*, 181
Dromidiopsis tridentata 67
Dromiidae 18, *67*, *73*, 134, 161
Dromioidea 18
Dungeness crab *see Cancer magister*
durian crab 50, *51*
Dynomenidae 12, 18, *35*, 50

E
echinoderms 30
Echinoecus pentagonus 122–3, *123*, *141*
ecological roles 104, 105, 112, *113*, 115, 118
edible crab *see Cancer pagurus*
eggs *see* reproduction
Emerita analoga 26, 192
endangered species 25, 45, *182*, 198–9, *198*, *199*, 216
Eocacinus praecursor 30
Eocene 33
Eoprosopon klugi 30, *30*
Epialtidae 160, 167, 171
epibionts 168
epidermis 82
Episesarma versicolor 181
epistomes *63*, *65*
Epixanthus dentatus 104–5, *104*
Erimacrus isenbeckii 187, 201, 211
Eriocheir 153
   E. japonica 191
   E. sinensis 189, *190*, 191, 196, 212, *213*
Eriphia 114
   E. sebana 114
Eriphiidae 181
estivation 15, 154
Ethusidae 116

Ethusina 116
   abyssicola 116
Eubrachyura 14–15, 18–19, 32, 35, *72*, *73*, 90, 144
Eumedoninae 141
eumedonine crabs 123
European edible crab *see Cancer pagurus*
European green crab *see Carcinus maenas*
evolution 14–15, 22–3, 25, 30–43, 138
   convergent 36–7
   fossil record 30–3, 138
   marine biographical regions 38
   morphological divergence 37
   parallel development 50
speciation 38–9
exophthalmy 101
exoskeleton 22, 31, 62, 82
   moulting 22, 75, *75*
eyes *63*, *74*, 74, *77*, 129
eyestalks 37, *63*, 64, 74, *74*, *108*

F
Fabricius, Johan Christian 54
face-stripe crab, Malaysian 98, *99*
false spider crab 12, 75, 149, 151
fiddler crab *43*, 68, 82, *83*, *108*, 144, 145, *145*, 156–7, *156*, 172, 181
   estuary *174*, 175
   thick-legged *174*, 175
filter feeders 26, 30, 104
fire crab 197
fishing industry 7, 180–9
   aquaculture 189
   boutique fisheries 184–5
   cultural food traditions 181–2
   introduced species 187, 196–7, 203
   major fisheries 183
   over-exploitation 25, *182*, 186, 188, 203, 211, 215
   sustainable practice 204
flagellum 64
flat coral crab 96, *97*
flower crab *see Portunus pelagicus*
flower moon crab 67
forceps crab 104–5, *104*
fossil record 30–3, 138
freshwater species 118, 153–4, 199
frog crab *14*, 53
   red *see Ranina ranina*
   slender *52*, 53

G
Galenidae 19
gall crab 57, 121
   green-spotted *56*, 57
Gandalfus yunohana 74
Garfunkel's crab 176, *177*
Garthambrus cidaris 67
gastric mill 65, 76, 101, 157
gaudy clown crab *88*, 89
Gazami crab 183, *183*
Gecarcinidae 19, 76, 119, 147, 164, 181, *201*
Gecarcinucidae 80, 118, 181
Gecarcinus ruricola 147, 181
Gecarcoidea natalis 16, *16–17*, 76, *107*, 147, *147*, *152–3*, 153, 164, *164*, 200–1, *200*
Gelasimus vocans 83
Geosesarma 118–19, *119*
   G. hagen 216, *217*
   G. notophorum 154, *154*
Geryonidae 19, 184
ghost crab 69, 101, 105, *107*, 149, 156–7
   Atlantic 69
   horn-eyed *100*, 101, 110
   painted *158*
   swamp 181, *181*
giant crab *see Pseudocarcinus gigas*
gills 22, 53, 65, 76, 77, 79, 80, *80*, 112
Glyptograpsidae 19, *67*
golden deep-sea crab 184, *184*
Gondwana 32
Goneplacidae 73, 119
Goniopsis cruentata 74
gonopods 71, 73, *73*, 144
gonopores 72, *72*, *73*
Grapsidae 19, *73*, 74, 78, 97, 110, 114, 130, 151, 181
Grapsoidea 19
Grapsus
   G. grapsus 73, 97, 110, *110*, 114, 130, *131*
   G. tenuicrustatus 114
green crab, European *see Carcinus maenas*
green gland 64
growth 22, 75, 116
guard crab *36*, 115
   honeycomb 121
Guinusia dentipes 67

H
haemolymph 76, 78, 80
horsehair crab *see Erimacrus isenbeckii*
hairy crab 54, *55*
Halimeda crab *see Huenia heraldica*
Hapalocarcinus marsupialis 57, 121

Hapalonotus pinnotheroides 123, 141
harlequin crab *see Lissocarcinus laevis*
Harryplax severus 125
heart *77*, 78, *78*, 79
Heloeciidae 19
helmet crab 144
Hemigrapsus
   H. sanguineus 195
   H. takanoi 195
hemocyanin 78
hepatic *63*
hermit crab 26, 28, 35, 46
   cone shell 46, *47*
   halloween 46, *47*
   hermit 29
   king 29, *29*
   orange-legged 46, *47*
   pink-clawed 28
   robber crab 28, 144
   strawberry 45
   striped 46, *47*
heterochelous chelae 68
Heterotremata 18, 35
Hexaplax aurantium 69
Hexapoda 22
Hexapodidae 12, *69*
Hippidae 26
Hippocrates 8
Hippoidea 26
Hoff crab 117, *117*
Homolidae 18
Homolodromiidae 18, 30
Homolodromioidea 18
Homoloidea 18
Hoplophrys oatesii 166, *167*
horned crinoid crab 120
horse crab 183
horsehair crab *see Erimacrus isenbeckii*
horseshoe crab *see Limulus polyphemus*
Hourdezi's hydrothermal vent crab *128*, 129
Huenia heraldica 160, *170*, 171
human diet, crab in 180–9, *180*, *181*, *182*, *183*, *186*, *187*, *188*, *189*
Hyasteneus 160
hydrothermal vents 15, 74, *74*, *107*, 107, 117, *128*, 129
Hymenosomatidae 75, 119, 149, 151
hypodermis 75
Hypothalassia
   H. acerba 188
   H. armata 188

I
Inachidae 116, 168
Inachoididae 85, 197
insects 22
intelligence *81*, 156–7

invasive species 194–7, 200, 203, 212
invertebrates 10
ion regulation 76
Iphiculidae 67
Iphiculus spongiosus 67
iridescent species 82
Ixa inermis 62

J
Johngarthia lagostoma 147
Jurassic 15, 30, 32–3

K
K-selected species 116, 151
Kamchatka crab *see Paralithodes camtschaticus*
king crab 29, 35, 180, 186–7
   Alaskan *see Paralithodes camtschaticus*
   blue 186, 187
   golden 186, 187
   king hermit crab 29, *29*
   Puget Sound *48*, 49
   red *see Paralithodes camtschaticus*
Kiwa 117
   K. hirsuta 117
   K. tyleri 117, *117*
krill 22

L
Labuanium
   L. trapezoideum 73
   L. vitatum 198
land crab 147
   blue 181
   brown *149*
   purple 181
larvae *see* reproduction (zoea)
Latreille, Pierre 34, *34*
Latreillia metanesa 33
Laurasia 32
lecithotrophy 153
legs *see* pereiopods
Leptodius exaratus 115
Leptograpsodidae 19
Leptomithrax gaimardii 126, *127*, *146*, 147
Leptuca pugilator 156
Leucosia anatum 7
Leucosiidae 67, 75
Lewindromia unidentata 134, *135*
life cycle of crabs *150*
Limulidae 25
Limulus polyphemus 22, *24–5*, 25
Linnaeoxanthidae 19
Linnaeus, Carl 11, 130
Liocarcinus puber 76
Liomera nigrimanus 67

Lissocarcinus
  L. laevis 123, *136*, 137, *148*
  L. orbicularis 123
Lithodes
  L. aequispina 192
  L. aequispinus 186
Lithodidae 29, 49, 180, 203
lobsters 22
Lomis hirta 29, *29*
Lomisoidea 29
longevity 116
Lopholithodes mandtii 48, *49*
Lophozozymus pulchellus *92*, 93
lopsided crab 90, *91*
luminescent species *82*, *92*, 93
lung fluke 190–1, *190*
Lupocyclus philippinensis 67
Lybia 66, 121
  L. leptochelis 86, *87*
  L. tessellata 15, 67
Lyreididae 18, 53
Lysirude 52, 53
  L. channeri 53

M

Mclaydromia dubia 161
Macrocheira kaempferi 12, 22, 23, 75
Macrophthalmidae 19, 67
Macrophthalmus 181
  M. abbreviatus 181
  M. grandidieri 67
Maja
  M. brachydactyla 127, 147, 184, 208, *209*
  M. squinado 184, 208
Majidae 126, 151, 208
Majoidea 160, 167, 171
mandarin crab 154, *154*
mandibles 64, 65, 76, 77
mangrove-dwellers 7, 15, 80, 82, 98, *99*, 104–5, *104*, 108, 111, *111*, 112, *113*, 156, 159
  mangrove re-oxygenation 80, 99, 104, 112
mangrove root crab 74
mangrove shore crab 158
manicou crab 105
manus 66, *66*
marine biographical regions *38*
marine environments *116*, 116–17
marsupial crab *see* *Hapalocarcinus marsupialis*
masking crab 160
Matuta planipes 67
Matutidae 18, 65, 67
maxillae 64, 65, 76
maxillipeds 65, 67, 76, 77
megalopae 151, *151*, 153, 189

memory *81*, 156
Menippe
  M. adina 184
  M. mercenaria 184
Menippidae 181
merus *63*, 66
mesogastric *63*
Mesozoic Marine Revolution 30–1
Metacarcinus magister 183
Metadynomene tanensis 35
Metaplax gocongensis 182
Metopaulias depressus 15, 37, 118, 154–5, *154*
Metopograpsus frontalis 110
Mictyridae 19, 133
Mictyris longicarpus 110, *132*, 133
migrations
  breeding 16, *16–17*, 146, 147, 164, 181, 200–1, *200*, 212
  Eriocheir 153
  Gecarcoidea natalis 16, *16–17*, 147, *147*, 152–3, 153, 164, *165*, 200–1, *200*
  larval 152–3, 153, 164, *165*
  Leptomithrax gaimardii 126, *127*, 146, 147
mimicry *159*, 160, *160*, *166*, 167, *170*, 171
mitten crab 153, 191
  Chinese *see* *Eriocheir sinensis*
mole crab 26, *26*, 192
monophyletic 36
mop crab 54, *55*
moulting 22, 75, *75*, 81, 127, 144, 147, 151
movement 12, 65, 68–9, *70*
  soldier crabs 133
  speed 12, 69, 101
mud crab
  Atlantic 106, *106*
  giant *see* *Scylla serrata*
  orange *182*
mud-dwellers 112, *113*
mutualism 120
Myriapoda 22

N

natatory legs *67*, 137
nauplius 22
nautiloids 30
Necora puber 109
Neorhynchoplax 149
  N. minima 12
Neosarmatium
  N. australiense 112
  N. integrum 78
nervous system 81, *81*
Notopus dorsipes 14

O

Oates's soft coral crab *166*, 167
Ocypode 69, *69*, 105, 149, 157
  O. ceratophthalmus 100, 101, 110
  O. cursor 105
  O. gaudichaudii 158
  O. quadrata 69, 157
  O. saratan 73
Ocypodidae 19, 67, 73, 101, 144, 156–7
Ocypodoidea 19
okadaic acid 192, 193
Ommatocarcinus minabensis 37
onchocerciasis 191, *191*
Opusia indica 73
orangutan crab 10
Oregoniidae 197
Orithyia sinica 184, *206*, 207
Orithyiidae 207
osmoregulation 80
ostia 78
Ovalipes
  O. catharus 157
  O. molleri 82
oxygen 78, 80
  mangrove re-oxygenation 80, 99, 104, 112
  pleonal flapping 148
Oziidae 181

P

paddle crab, New Zealand 157
Palaeocene 33
palaeogeography 40
Palaeozoic 30
Palicidae 12
palytoxin 192, 193
Pangaea 32
Panopeidae 19
Panopeus herbstii 106
Paradorippe
  P. australiensis 161
  P. granulata 161
paragonimiasis 190–1, *190*
Paraleptuca crassipes 83, *174*, 175
Paralithodes 192
  P. camtschaticus 49, 186–7, *196*, *202*, 203
  P. platypus 186
Paramedaeus megagomphios 67
Parasesarma 98
  P. eumolpe 158
  P. leptosoma 156
  P. peninsulare 98, *99*
parasitic species 141
Parthenopidae 19, 67
Parthenopoidea 19
Patagurus rex 29, *29*
pea crab 12, 120, *120*, 123

pebble crab 75
  ringed *see* *Leucosia anatum*
pentagonal sea urchin crab 122–3, 123
Percnidae 19, 67, 97
Percnon
  P. affine 67
  P. gibbesi 97
  P. guinotae 96, 97
pereiopods 12, 22, *63*, 65, 68–70, *70*
  carrier legs 12, 61
  natatory *67*, 137
  vestigial 12
Periacanthus horridus 32
periscope crab 181
Permian 30–1
petal-eyed swimming crab 37
Petrolisthes violaceus 27
phyllobranchiate gills 80
Phyllotymolinidae 18
phytotelmata *42*, 118–19
Pilumnidae 19, 54, 67, *120*, 123, 141
Pilumnoidea 19
Pilumnoididae 19
Pilumnus
  P. granti 67
  P. vespertilio 54, *55*
pincer *see* chela
pink carrier crab 94, *95*
Pinnixa 12
  P. tumida 120
Pinnotheres pisum 120
Pinnotheridae 12, 19, 120, 123
Pinnotheroidea 12, 19, 57
Plagusiidae 19, 67
plankton 32, 107, *107*
planktonic larval crabs 7, 151, 198
Planopilumnidae 19
Planes 39
plate crab 159
Platychirograpsus spectabilis 67
Platykotta akaina 32
Platypodiella spectabilis 88, 89
pleon 12, 26, 62, *63*, 71, *71*, 72
pleonal flapping 148
pleopods 71
Podocatactes hamifer 90, *91*
Podotremata 14–15, 18, 32, *33*, 34, 35, 72, *72*, 94, 144, 161
pollex *63*, 66, *66*
pollutants 78, *78*
polyphyletic species 36
pom-pom crab 15, 86, 87
porcelain crab 27, 35
  violet 27

porter crab 94, 161
  granulated 161
  leaf-porter 161
  mauve 161
  rough-shelled 161, 94, *95*
Portunidae 19, 34, 65, 67, 69, 104, 123, 137, 151, 183, 189
Portunoidea 19
Portunus
  P. armatus 40, 72
  P. pelagicus 40–1, *40*, 83, *183*, 189
  P. reticulatus 40
  P. trituberculatus 183, *183*
Potamidae 19, 181
Potamoidea 19
Potamon 8–9
  P. fluviatile 9, *9*
Potamonautes
  P. lirrangensis 191
  P. orbitospinus 118
Potamonautidae 18, 191
Potamonidae 118, 181
prawns 22
pretty xanthid crab 92, 93
prey, crabs as 108–11, *108*, *109*, *110*, 118, 158–9
  see also fishing industry
prickly dynomenid crab 50, *51*
Procambarus clarkii 118
propodus *63*
protogastric *63*
Pseudocarcinidae 19
Pseudocarcinoidea 19
Pseudocarcinus gigas 6, 12, 185, *185*
Pseudocryptochirus viridis 56, 57
Pseudorhombilidae 19
Pseudothelphusidae 19, 105
Pseudothelphusoidea 19
Pseudoziidae 19
Pseudozioidea 19
pterygostome *63*, 80
pycnogonids 22
Pyreneplax basaensis 138
Pyromaia tuberculata 197

Q

queen crab *see* *Carpilius corallinu*; *Erimacrus isenbeckii*

R

r-selected species 151
Ranina ranina 6, 67, 185, 204, *205*
Raninidae 14, 18, 67, 204
Raninoidea 18, 53
red crab *see* *Gecarcoidea natalis*
red deep-sea crab 187, 188
red devil crab 216, *217*

reef crab 7, *82*, 114, 115
  Adam's zebra crab *140*, 141
  black-fingered 193
  Guinot's agile *96*, 97
  pretty crested *92*, 93
  purple-ocellated *40*
  symbiosis 57, 115, *115*, 120, *120*, 160
reef rubble crab *159*
reproduction 15, 72–3, *72*, *73*
  abbreviated development 153–4
  active brood-care 15, 37, 58, 71, 148, *148*, 153–4
  allopatric speciation 39
  breeding migrations *146*, 147, *147*, 164, 181, 200–1, *200*, 212
  courtship 144–5, 156–8, *172*, *173*, *174*, 175
  egg incubation 145
  eggs 7, 15, 22, 58, 71, 72, *72*, 148–50, *149*, *150*, 151, 153
  external fertilization 14, 35, 72, *72*, 144
  gonopods 71, 73, *73*, 144
  gonopores 72, *72*, 73
  internal fertilization 14, 72, *72*, 73, 144
  K-selected species 116, 151
  larval migration 152–3, 153, 164, *165*
  lecithotrophy 153
  life cycle *150*
  mating 68, 98, 156–7, *184*
  megalopa 151, *151*, 153
  nauplius 22
  ovary and oviduct 72–3, *72*
  ovoviviparous species 149, 153–4
  pleopods 71
  r-selected species 151
  reproductive organs 35, 72–3, *72*, 73, 144
  sexual dimorphism 144
  sperm 72–3, *72*
  synchronized larval release 149
  viviparous species 15, 42, 118
  zoea (larvae) 22, 38, 58, 129, 147, 149, *150*, 151, *151*, 153, 155, *163*, 198
  zoeal development *150*, 151, *151*, 153
  zoeal dispersal 38, 151, 153
respiratory system 80, *80*
Retroplumidae 12, 19
Retroplumoidea 19
Ricinulei 25
river blindness 191

river crab 8–9
  Chinese *see Eriocheir sinensis*
  Mediterranean 9, *9*
robber crab *see Birgus latro*
rock crab 78
  Atlantic *see Cancer irroratus*
  red 193
  red rock *see Grapsus grapsus*
rock shore-dwellers 114, *114*, *115*, 130, *131*
Rodriguezus garmani 105
Rosse, William Parsons, 3rd Earl 10, *11*
rosy egg crab *194*
royal crab 188

S

Sakaila africana 67
Sally Lightfoot *see Grapsus grapsus*
sand bubbler crab *80*
sand crab 26, 192
  horn-eyed *100*, 101
  priest 27
sand-dwellers *132*, 133
Sarmatium unidentatum 67
saxitoxin 192–3
scaphognathite 65
scavengers 30, 32, 45, 104, 107, *107*, 193
Schizophrys aspera 115
Scopimera inflata 67, *80*
sculptured crab 138, *139*
Scylla 189
  S. olivacea 182
  S. serrata 7, 41, *41*, 76, *150*, 151, 183
sea anemones 13, 15, 85, 86, 123
  boxer crabs *15*, 66, 86, *87*, 121
  as camouflage *28*, 161
sea-urchin-carrying crab 94, *95*
senses 81
  chemicals, sensitivity to 90, 106
sentinel crab 181, *181*
Serenepilumnus kukenthali 75
Sesarmidae 19, 37, 67, 73, 76, 80, 82, 98, 112, 119, 176, 181, 216
Sesarmoides 37
setae 75, 80, 81, 133, *156*, 159
sexual dimorphism 12, 144
shag-pile crab 75
shaggy shore crab 54, *55*
shame-faced crab, common *see Calappa lophos*
shawl crab *see Atergatis floridus*
shell *see* carapace
shrimps 22

sight, sense of 74, *74*, 81, *108*
smell, sense of 64, 81, 106
snow crab *see Chionoecetes opilio*
soldier crab 110, 181
  East Australian *132*, 133
  red-kneed *132*, 133
somites 62, 71, *71*
Southern Crab Nebula 10
Southern kelp crab 7
spanner crab *see Ranina ranina*
speciation 38–9
spider crab *84*, 85, 116, *116*, 151, 160
  arrowhead crab *170*, 171
  common *see Maja brachydactyla*
  European *see Maja brachydactyla*
  Gaimard's *see Leptomithrax gaimardii*
  great (giant) *see Leptomithrax gaimardii*
  Japanese *see Macrocheira kaempferi*
  migrations *146*, 147
  Periacanthus horridus 32
  red 115
  spiny 184
  stalk-eyed *33*
spindle crab 62
spiny-clawed deep-sea crab 58, *59*
sponge crab 134, 161
  Lewinsohn's 134, *135*
  Mclaydromia dubia 161
Stalin's crab *see Paralithodes camtschaticus*
stalk-eyed crab 145
  stalk-eyed shore crab *172*, *173*
statocysts 64, *64*
Stenorhynchus seticornis 84, 85
stone crab 29, 180
  Florida 184, *184*
  Gulf 184
  hairy 29, *29*
stridulation 156–7, *157*, 158
suspension feeders 27
swimmerets 27, 71
swimming 65, 69, 104
swimming crab 151, 189
  green *109*
  harlequin *see Lissocarcinus laevis*
  Indo-West Pacific 156
  ridged 144
symbiosis 120–3, *120*, *122*, 123
  boxer crabs *15*, 66, 86, *87*, 121
  gall crabs 57
  Lissocarcinus 136, 137

mimicry *159*, 160
  obligate 120
  reef-dwellers 57, 115, *115*, 160
Symethis corallica 67

T

tagmata 62
Taliepus nuttallii 7
Tanaochelidae 19
tanner crab 147, 192, 193
  Baird's *see Chionoecetes bairdi*
Tasmanoplax latifrons 67
teeth 67, 76
tekson 63
Telmessus cheiragonus 144
telson 14, 63, 71, 76
terrestrial species 118–19, 149, 153
Tetralia glaberrima 73
Tetraliidae 19, *36*, 73, 115, 120
Tetraloides heterodactyla 36
tetrodotoxin 192–3
Thalamita 109
  T. sima 67
thoracic sternum 63, 72
Thoracotremata 19, 33, 35, 57
Thranita crenata 156
tiger crab *see Orithyia sinica*
tiger-faced crab *see Orithyia sinica*
Tmethypocoelis ceratophora 145, *172*, *173*
Tokoyo eburnea 67
toxic species 89, *92*, 93, 192–3, *192*, *193*
Trapezia 120
  T. cymodoce 36
  T. septata 121
Trapeziidae 19, *36*, 115, 120
Trapezioidea 19
tree-climbing species 118
Triassic 31–2
Triassic–Jurassic extinction event 30
Trichodactylidae 19, 118
Trichodactyloidea 19
Trichopeltariidae 90
trilobites 30
Trilobitomorpha 22
troglobites 119, *119*
troglobitic crabs *see* cave-dwelling species
Troglocarcinus 153
Trogloplacinae 153
Trogloplax joliveti 119, *119*
Tubuca
  T. coarctata 43
  T. flammula 83
  T. paradussumieri 83
  T. polita 145
Tumidodromia dormia 73

U

Uca 82, 144, 157, 175, 181
  U. dussumieri 67
Ucides cordatus 181, *181*
urchin zebra crab *140*, 141
urogastric 63
uropods 14

V

vampire crab 118–19, *119*
  little red 216, *217*
Varunidae 19, 80, 119, 159, 182, 196, 212
vectors, crabs as 190–1
velcro crab 168, *169*
velvet crab *109*
  orange *35*
vicariance 39
vinegar crab 181, *181*
viviparous species 15, 42, 118
Vultocinidae 138
Vultocinus anfractus 138, *139*

W

white-stripe crab *198*

X

Xanthasia murigera 123
Xanthias maculatus 40
Xanthidae 19, 67, 86, 89, 93, 115, 120, 121, 181
Xanthoidea 19, 193
Xenocarcinus 160
  X. conicus 159
Xenograpsidae 19
Xenograpsus testudinatus 107, *107*
Xenophthalmidae 19
Xiphosura 22, 25

Y

yellowline arrow crab *84*, 85
Yeti crab 117
Yuebeipotamon calciatile 199

Z

Zebrida adamsii 140, 141
zoea *see* reproduction
zoonosis, food-borne 190–1, *190*
Zosimus aeneus 193

PICTURE CREDITS

Cover details
Front: itor/Shutterstock
Spine: Tin-Yam Chan

Agefotostock/ Colin Marshall 15

Alamy Stock Photo/ Alessandro Mancini 181tr; Amadeu Ito 181c; Andrey Nekrasov 202; blickwinkel 68, 174; Blue Planet Archive 76; David Fleetham 122-123; David Kleyn 201; Design Pics Inc 180; Eitan Simanor 189; Felix Lipov 183b; FLPA 154l; Frank Hecker 213t; George Grall 105; Helmut Corneli 24-25, 118; Image Professionals GmbH 165; Imagebroker 100, 136; Imaginechina Limited 190l; James Peake 146; Kelvin Aitken / VWPics 33t; Michael Patrick O'Neill 48, 184r; Michelle Gilders 131; Minden Pictures 16-17, 44, 107tl, 185, 200t; Natalia Kuzmina 69t; Nataly Mayak 187t; Nature Photographers Ltd 155; Nature Picture Library 119t, 120t, 127, 158, 192; Oksana Maksymova 56; Our Wild Life Photography 109tr; Paulo Oliveira 213b; Perry van Munster 65l; Premier 28; Reinhard Dirscherl 169; Stocktrek Images, Inc. 84; WaterFrame 66, 152-153, 214

Father Alejandro J. Sánchez Muñoz 154r

Alev Ozten Low 111t

Alexander Semenov/Aquatilis 210

Andrey Ryanskiy 7l, 13, 42t, 92t

Antonio De Angeli, from De Angeli & Caporiondo, 2009 32

Arthur Anker 12, 27t, 29t, 47, 55, 83bl, 151, 156

Brendan Schembri 111b

Carrie Schweitzer 30

Chao Huang 199

Courtesy of Charles Fisher/The Woods Hole Oceanographic Institution 128

Chris Bray - Swell Lodge 198

Ellen Muller/pbase.com/imagine 89, 115r

FLPA/ Steve Trewhella 39

Getty Images/ Ken Usami 23; patti white photography 109b; Ricardo Lima 181b; Westend61 2

Hsi-Te Shih 83tl, 83tr, 83tcr, 83bcl, 83bcr, 173

Iain McGaw 79

Ian Banks 72

Ian Kanda (iNaturalist:Crustaceans/CC 1.0) 106

Ian Shaw 96

Internet Archive 34

iStock/ Jake Davies 209; scubaluna 170

© Jan van de Kam, Netherlands 83tcl, 83br, 112, 145

Kevin Lee 7r

M. Juncker 41 (courtesy of Muséum national d'Histoire naturelle)

Marc Damant/©Damant's Diving Digital Photography 166

Mark Norman/Museums Victoria/CC BY 29b

Max Orchard 177

NASA, ESA, J. Hester and A. Loll (Arizona State University) 11

U.S. National Oceanographic and Atmospheric Office of Ocean Exploration and Research 116, 184l, **/CC BY-SA 2.0** 187b

© Karen Gowlett-Holmes CSIRO Oceans & Atmosphere– the NORFANZ Expedition 82l

Lee Kong Chian Natural History Museum, Singapore
Paul Ng 42b, 75r, 107b; Peter Ng 51, 74b, 119b, 139, 206

Naturepl.com/ Gary Bell/Oceanwide 6b; Emanuele Biggi 9; Sue Daly 64; Brandon Cole 141

Oliver Mengedoht/Panzerwelten.de 191b

Ondřej Radosta/www.crabdatabase.info 65r, 159t, 183t

Peter Davie/Queensland Government 150, 182tl, 182tr

Queensland Museum 80l, 150; Jeff Wright 43, 80r; Neville Coleman Collection 36r, 104, 161l; / National University of Singapore 40b

© Roger Steene 5, 6t, 36l, 37, 62, 75l, 78l, 82r, 92b, 115l, 120b, 123tr, 135, 144, 148, 149l, 149r, 159b, 160, 161r, 162, 193, 205

Russell Constable 27b

Shutterstock/ alonanola 10; Ammit Jack 74t; fluke samed 78r; Marut Sayannikroth 99; Red ivory 109tl; Steven Grogger 110t; Yai 110bl; almonod 110br; pryg123 114t; divedog 117 b/g; Ken Griffiths 132; nattpat.mos 181tl; Pooh and ball 182b; Shpatak 186; oakemon 188; Medtech THAI STUDIO LAB 249 190c; Levent Konuk 194; Anastasia_Fisechko 196; Kondratuk Aleksei 197; Dan Olsen 217

SWNS 70

Tan Heok Hui (heokhui@nus.edu.sg) 95, 114b, 107tr

Tin-Yam Chan, National Taiwan Ocean University/Muséum national d'Histoire naturelle 3, 14, 33b, 35, 40t, 52, 59, 69b, 71, 91

© 2015 Thatje et al. Thatje S, Marsh L, Roterman CN, Mavrogordato MN, Linse, K (2015). Adaptations to Hydrothermal Vent Life in Kiwa tyleri, a New Species of Yeti Crab from the East Scotia Ridge, Antarctica. PLoS ONE. 10(6): e0127621. doi:10.1371/journal.pone.0127621/CC BY 117

Unsplash.com/raphael bick 147

Uwe Weinreich 195

Wikipedia Commons/ *Emerita analoga*/jkirkhart35, CC BY 2.0 26; *Callichimaera perplexa* Luque et al 2019/Image by Oksana Vernygora, supplied by Javier Luque who commissioned the drawing, CC BY-SA 4.0 31; *Paragonimus westermani*/Division of Parasitic Diseases and Malaria 190r; *Onchocerca volvulus* emerging from a black fly/United States Department of Agriculture 191t

Yoshihisa Fujita (Okinawa Prefectural University of Arts) 121, 124

Yisrael Schnitzer (Bar-Ilan University) 87

Yvonne McKenzie, Wondrous World Images 200b

ILLUSTRATIONS

Major diagrams were newly prepared for this volume based on a variety of sources listed below. The only exceptions are some detailed scientific drawings that appear in the Classification Table (pp18–19), or as examples of claw and gonopod types (pp67, 73). These were either by the present author; out of copyright; the scientists are deceased and can no longer be contacted; or used with permission. Key sources were: p38 (after Ocean Biodiversity Information System, obis.org/2018/11/30/adam); p64 (after Warner 1977, Sandeman & Okajima 1972); p67 (variously after Chen 1989, Chen & Sun 2002, Crosnier 1962 & 1965, Davie 1989, 1992 & 1997, Garth & Davie 1994, George & Jones 1982, Griffin 1969 & 1973, Guinot 1976, Kemp 1915, Manning & Holthuis 1981, McCulloch & McNeill 1923, McLay 1993, Montgomery 1931, Ng 1998, Ng & Tan 1984, Rathbun 1918, Tavares 1993, Tesch 1915); p70 (after Vidal-Gadea & Belanger 2009); pp71, 73tr (after Ng 1998); p72 (after Hartnoll 1968a, Guinot 1978); pp77, 81 (after Pearson 1908, Felgenhauer 1992); p108 (after Zeil & Hemmi 2010); p145 (after Davie & Kosuge 1995); p157 (after Boon, Yeo & Todd 2009, Guinot-Dumortier & Dumortier 1960). Full citations for these are listed in a scientific review by Davie, Guinot and Ng (2015) – see Chapter 71-2, 'Anatomy and Functional Morphology of Brachyura', pp11–163, in Castro et al. (2015), cited in Further Reading on p218.

ACKNOWLEDGEMENTS

This book is the result of a lifetime's fascination with the marine world, and of course with crabs in particular. As such, I owe thanks and appreciation to everyone who has accompanied, supported, and encouraged me on this journey. Eddie Hegerl, one of Australia's leading marine conservationists, infected me with his enthusiasm for mangrove and reef protection while I was still in high school, encouraging me to become the 'crab expert' on Queensland Littoral Society surveys of mangrove swamps. Bruce Campbell of the Queensland Museum became my crustacean taxonomic mentor, and found a job for me from which I never left!

Of my crustacean colleagues, I have had wonderful support from so many great people. The 'Paris Museum group' in the 1990s, especially Alain Crosnier and Danièle Guinot, became friends and mentors. Alain found the funds that enabled extended stays at the Muséum national d'Histoire naturelle. Danièle is a legendary carcinologist and it is a privilege to continue to collaborate with her. My long-time friend the late Michael Türkay of the Senkenberg Museum, Frankfurt, was a kindred spirit and a great font of knowledge who taught me much. Peter Ng, of the National University of Singapore, has been my longest and closest professional colleague and personal friend, and in a way, we have grown up and grown old together, sharing our great abiding passion. My debt to him is immeasurable, and many of the ideas in this book evolved through our long talks together.

My old friend Roger Steene, a remarkable Australian underwater photographer, has for many years shared my excitement with crabs and generously provided many images for this book. Numerous colleagues have also provided fabulous images, and I would particularly like to thank Arthur Anker, Tin-Yam Chan, Yoshihisa Fujita, Jan van de Kam, Mike McCoy, Peter Ng, Max Orchard, Ondřej Radosta, Andrey Ryanskiy, Hsi-Te Shih and the Queensland Museum. Others also offered great images that could not be used because of space, but my gratitude goes to all of you for your generosity. Paul Clark, Miranda Lowe and Kevin Webb of the Natural History Museum, London, went above and beyond to secure some lovely old illustrations of crab anatomy. Dirk Brandis gave me some wonderful insights into the biological origins behind the astrological symbology of Cancer. Carrie Schweitzer and Rod Felder shared a photograph of a fossil crab, and gave helpful feedback on the evolutionary history text. A lovely formal review of the manuscript by Patricia Backwell led to some real improvements. A big thank you to Marissa McNamara of the Queensland Museum for her careful reading and editing of my original manuscript, and making lots of great suggestions. Darryl Potter of the QM also deserves a special mention for standing by my side as a collection manager and field/dive partner for over 35 years – so many shared carcinological experiences! Finally, a special thank you to my friend the inimitable Lisa-Ann Gershwin, whose beautiful *Jellyfish: A Natural History* was my inspiration for a similar book on crabs; and more practically for first introducing me to the team at Ivy Press.

Several people have been instrumental in seeing *Crabs* grow from a nice idea into a beautiful book. Senior commissioning editor Kate Shanahan was my initial contact at Ivy Press, and she, and Natalia Price-Cabrera who followed her, were invaluable in helping me turn it into a viable project. Abbie Sharman, development editor at The Bright Press, has been wonderfully supportive. It has been a real pleasure working with Angela Koo as the project editor, and with the hugely creative Jane Lanaway designing the layouts. I could not have asked for a better group of people to see my book through to fruition. Finally, a big thank you to Princeton University Press for embracing *Crabs* and ensuring its ultimate birth. Thank you all.

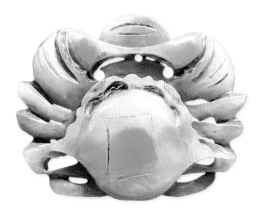

The Chinese characters for crab (蟹) and harmony (鰓) are both pronounced *xie,* and thus crab charms can symbolize a desire for peace. When the crab has a tight hold on a coin it also represents prosperity. May the spirit of the crab bring peace and prosperity into all our lives.